Reviews of Environmental Contamination and Toxicology

Continuation of Residue Reviews

VOLUME 238

More information about this series at http://www.springer.com/series/398

Reviews of Environmental Contamination and Toxicology

Editor
Pim de Voogt

Volume 238

Springer

ISSN 0179-5953 ISSN 2197-6554 (electronic)
Reviews of Environmental Contamination and Toxicology
ISBN 978-3-319-30790-9 ISBN 978-3-319-30791-6 (eBook)
DOI 10.1007/978-3-319-30791-6

Printed on acid-free paper

This Springer imprint is published by Springer Nature
The registered company is Springer International Publishing AG Switzerland

In memoriam David M. Whitacre

Foreword

International concern in scientific, industrial, and governmental communities over traces of xenobiotics in foods and in both abiotic and biotic environments has justified the present triumvirate of specialized publications in this field: comprehensive reviews, rapidly published research papers and progress reports, and archival documentations These three international publications are integrated and scheduled to provide the coherency essential for nonduplicative and current progress in a field as dynamic and complex as environmental contamination and toxicology. This series is reserved exclusively for the diversified literature on "toxic" chemicals in our food, our feeds, our homes, recreational and working surroundings, our domestic animals, our wildlife, and ourselves. Tremendous efforts worldwide have been mobilized to evaluate the nature, presence, magnitude, fate, and toxicology of the chemicals loosed upon the Earth. Among the sequelae of this broad new emphasis is an undeniable need for an articulated set of authoritative publications, where one can find the latest important world literature produced by these emerging areas of science together with documentation of pertinent ancillary legislation.

Research directors and legislative or administrative advisers do not have the time to scan the escalating number of technical publications that may contain articles important to current responsibility. Rather, these individuals need the background provided by detailed reviews and the assurance that the latest information is made available to them, all with minimal literature searching. Similarly, the scientist assigned or attracted to a new problem is required to glean all literature pertinent to the task, to publish new developments or important new experimental details quickly, to inform others of findings that might alter their own efforts, and eventually to publish all his/her supporting data and conclusions for archival purposes.

In the fields of environmental contamination and toxicology, the sum of these concerns and responsibilities is decisively addressed by the uniform, encompassing, and timely publication format of the Springer triumvirate:

Reviews of Environmental Contamination and Toxicology [Vol. 1 through 97 (1962–1986) as Residue Reviews] for detailed review articles concerned with any aspects of chemical contaminants, including pesticides, in the total environment with toxicological considerations and consequences.

Bulletin of Environmental Contamination and Toxicology (Vol. 1 in 1966) for rapid publication of short reports of significant advances and discoveries in the fields of air, soil, water, and food contamination and pollution as well as methodology and other disciplines concerned with the introduction, presence, and effects of toxicants in the total environment.

Archives of Environmental Contamination and Toxicology (Vol. 1 in 1973) for important complete articles emphasizing and describing original experimental or theoretical research work pertaining to the scientific aspects of chemical contaminants in the environment.

The individual editors of these three publications comprise the joint Coordinating Board of Editors with referral within the board of manuscripts submitted to one publication but deemed by major emphasis or length more suitable for one of the others.

Coordinating Board of Editors

Preface

The role of *Reviews* is to publish detailed scientific review articles on all aspects of environmental contamination and associated (eco)toxicological consequences. Such articles facilitate the often complex task of accessing and interpreting cogent scientific data within the confines of one or more closely related research fields.

In the 50+ years since *Reviews of Environmental Contamination and Toxicology* (formerly *Residue Reviews)* was first published, the number, scope, and complexity of environmental pollution incidents have grown unabated. During this entire period, the emphasis has been on publishing articles that address the presence and toxicity of environmental contaminants. New research is published each year on a myriad of environmental pollution issues facing people worldwide. This fact, and the routine discovery and reporting of emerging contaminants and new environmental contamination cases, creates an increasingly important function for *Reviews*. The staggering volume of scientific literature demands remedy by which data can be synthesized and made available to readers in an abridged form. *Reviews* addresses this need and provides detailed reviews worldwide to key scientists and science or policy administrators, whether employed by government, universities, nongovernmental organizations, or the private sector.

There is a panoply of environmental issues and concerns on which many scientists have focused their research in past years. The scope of this list is quite broad, encompassing environmental events globally that affect marine and terrestrial ecosystems; biotic and abiotic environments; impacts on plants, humans, and wildlife; and pollutants, both chemical and radioactive; as well as the ravages of environmental disease in virtually all environmental media (soil, water, air). New or enhanced safety and environmental concerns have emerged in the last decade to be added to incidents covered by the media, studied by scientists, and addressed by governmental and private institutions. Among these are events so striking that they are creating a paradigm shift. Two in particular are at the center of ever increasing media as well as scientific attention: bioterrorism and global warming. Unfortunately, these very worrisome issues are now superimposed on the already extensive list of ongoing environmental challenges.

The ultimate role of publishing scientific environmental research is to enhance understanding of the environment in ways that allow the public to be better informed or, in other words, to enable the public to have access to sufficient information. Because the public gets most of its information on science and technology from internet, TV news, and reports, the role for scientists as interpreters and brokers of scientific information to the public will grow rather than diminish. Environmentalism is an important global political force, resulting in the emergence of multinational consortia to control pollution and the evolution of the environmental ethic. Will the new politics of the twenty-first century involve a consortium of technologists and environmentalists, or a progressive confrontation? These matters are of genuine concern to governmental agencies and legislative bodies around the world.

For those who make the decisions about how our planet is managed, there is an ongoing need for continual surveillance and intelligent controls to avoid endangering the environment, public health, and wildlife. Ensuring safety-in-use of the many chemicals involved in our highly industrialized culture is a dynamic challenge, because the old, established materials are continually being displaced by newly developed molecules more acceptable to federal and state regulatory agencies, public health officials, and environmentalists. New legislation that will deal in an appropriate manner with this challenge is currently in the making or has been implemented recently, such as the REACH legislation in Europe. These regulations demand scientifically sound and documented dossiers on new chemicals.

Reviews publishes synoptic articles designed to treat the presence, fate, and, if possible, the safety of xenobiotics in any segment of the environment. These reviews can be either general or specific, but properly lie in the domains of analytical chemistry and its methodology, biochemistry, human and animal medicine, legislation, pharmacology, physiology, (eco)toxicology, and regulation. Certain affairs in food technology concerned specifically with pesticide and other food-additive problems may also be appropriate.

Because manuscripts are published in the order in which they are received in final form, it may seem that some important aspects have been neglected at times. However, these apparent omissions are recognized, and pertinent manuscripts are likely in preparation or planned. The field is so very large and the interests in it are so varied that the editor and the editorial board earnestly solicit authors and suggestions of underrepresented topics to make this international book series yet more useful and worthwhile.

Justification for the preparation of any review for this book series is that it deals with some aspect of the many real problems arising from the presence of anthropogenic chemicals in our surroundings. Thus, manuscripts may encompass case studies from any country. Additionally, chemical contamination in any manner of air, water, soil, or plant or animal life is within these objectives and their scope.

Manuscripts are often contributed by invitation. However, nominations for new topics or topics in areas that are rapidly advancing are welcome. Preliminary communication with the Editor-in-Chief is recommended before volunteered review manuscripts are submitted. *Reviews* is registered in WebofScience™.

Inclusion in the Science Citation Index serves to encourage scientists in academia to contribute to the series. The impact factor in recent years has increased from 2.5 in 2009 to almost 4 in 2013. The Editor-in-Chief and the Editorial Board strive for a further increase of the journal impact factor by actively inviting authors to submit manuscripts.

Amsterdam, The Netherlands
January 2015

Pim de Voogt

Contents

Contributors

Alessia Carocci Dipartimento di Farmacia-Scienze del Farmaco, Universitá degli Studi di Bari "A. Moro", Bari, Italy

Alessia Catalano Dipartimento di Farmacia-Scienze del Farmaco, Universitá degli Studi di Bari "A. Moro", Bari, Italy

Samuel Dampare National Radioactive Waste Management Centre, National Nuclear Research Institute, Ghana Atomic Energy Commission, Accra, Ghana

Lisette N. De Senerpont Domis Department of Aquatic Ecology, Netherlands Institute of Ecology (NIOO-KNAW), Wageningen, Netherlands

Aquatic Ecology and Water Quality Management Group, Department of Environmental Sciences, Wageningen University, Wageningen, Netherlands

Ellen Van Donk Department of Aquatic Ecology, Netherlands Institute of Ecology (NIOO-KNAW), Wageningen, Netherlands

Department of Ecology and Biodiversity, University of Utrecht, Utrecht, Netherlands

Israel Doyi National Radioactive Waste Management Centre, National Nuclear Research Institute, Ghana Atomic Energy Commission, Accra, Ghana

David Kofi Essumang Graduate School of Nuclear and Allied Sciences, Ghana Atomic Energy Commission, Accra, Ghana

Giuseppe Genchi Dipartimento di Farmacia e Scienze della Salute e della Nutrizione, Universitá della Calabria, Cosenza, Italy

Eric Tetteh Glover Department of Chemistry, University of Cape Coast, Cape Coast, Ghana

Katharina Grosser Department of Aquatic Ecology, Netherlands Institute of Ecology (NIOO-KNAW), Wageningen, Netherlands

German Centre for Integrative Biodiversity Research (iDiv) Halle-Jena-Leipzig, Leipzig, Germany

Graziantonio Lauria Dipartimento di Farmacia e Scienze della Salute e della Nutrizione, Universitá della Calabria, Cosenza, Italy

Miquel Lürling Department of Aquatic Ecology, Netherlands Institute of Ecology (NIOO-KNAW), Wageningen, Netherlands

Aquatic Ecology and Water Quality Management Group, Department of Environmental Sciences, Wageningen University, Wageningen, Netherlands

Sara Mostafalou Department of Pharmacology and Toxicology, School of Pharmacy, Ardabil University of Medical Sciences, Ardabil, Iran

Scott Peacor Department of Aquatic Ecology, Netherlands Institute of Ecology (NIOO-KNAW), Wageningen, Netherlands

Department of Fisheries and Wildlife, Michigan State University, East Lansing, MI, USA

Ana Rita Ribeiro LCM – Laboratory of Catalysis and Materials – Associate Laboratory LSRE/LCM, Faculdade de Engenharia, Universidade do Porto, Porto, Portugal

Cláudia Ribeiro CESPU, Instituto de Investigação e Formação Avançada em Ciências e Tecnologias da Saúde (IINFACTS), Rua Central de Gandra, Gandra, PRD, Portugal

Maria Stefania Sinicropi Dipartimento di Farmacia e Scienze della Salute e della Nutrizione, Universitá della Calabria, Cosenza, Italy

Maria Elizabeth Tiritan CESPU, Instituto de Investigação e Formação Avançada em Ciências e Tecnologias da Saúde (IINFACTS), Rua Central de Gandra, Gandra, PRD, Portugal

Laboratório de Química Orgânica e Farmaceûtica, Departamento de Ciências Químicas, Faculdade de Farma´cia, Universidade do Porto, Porto, Portugal

Centro Interdisciplinar de Investigação Marinha e Ambiental (CIIMAR/CIMAR), Universidade do Porto, Porto, Portugal

Priority Substances and Emerging Organic Pollutants in Portuguese Aquatic Environment: A Review

Cláudia Ribeiro*, Ana Rita Ribeiro*, and Maria Elizabeth Tiritan

Contents

---

*These authors equally contributed to this work.

C. Ribeiro (✉)
CESPU, Instituto de Investigação e Formação Avançada em Ciências e Tecnologias da Saúde (IINFACTS), Rua Central de Gandra, 1317, 4585-116 Gandra PRD, Portugal
e-mail: claudia.ribeiro@iscsn.cespu.pt

A.R. Ribeiro (✉)
LCM – Laboratory of Catalysis and Materials – Associate Laboratory LSRE/LCM, Faculdade de Engenharia, Universidade do Porto, Rua Dr. Roberto Frias, 4200-465 Porto, Portugal
e-mail: ritalado@fe.up.pt

M.E. Tiritan
CESPU, Instituto de Investigação e Formação Avançada em Ciências e Tecnologias da Saúde (IINFACTS), Rua Central de Gandra, 1317, 4585-116 Gandra PRD, Portugal

Laboratório de Química Orgânica e Farmacêutica, Departamento de Ciências Químicas, Faculdade de Farmácia, Universidade do Porto

Centro Interdisciplinar de Investigação Marinha e Ambiental (CIIMAR/CIMAR), Universidade do Porto, Rua dos Bragas 289, 4050-123 Porto, Portugal

P. de Voogt (ed.), *Reviews of Environmental Contamination and Toxicology, Volume 238*, Reviews of Environmental Contamination and Toxicology 238,
DOI 10.1007/398_2015_5006

1 Introduction

Aquatic environments are among the most noteworthy ecosystems regarding chemical pollution due to the anthropogenic pressure. In 2000, the European Commission implemented the Water Framework Directive (WFD), with the aim of progressively reducing aquatic chemical pollution of the European Union (EU) countries. Therefore, the knowledge about the chemical and ecological status is imperative to determine the overall quality of water bodies. Concerning Portugal, some studies have demonstrated the presence of pollutants in the aquatic environment but an overall report is not available yet. The aim of this paper is to provide a comprehensive review about the occurrence of Priority Substances (PSs) included in the WFD and some classes of Emerging Organic Pollutants (EOPs) that have been found in Portuguese aquatic environment. The most frequently studied compounds comprise industrial compounds, natural and synthetic estrogens, phytoestrogens, phytosterols, pesticides, pharmaceuticals and personal care products. Concentration of these pollutants ranged from few ng L^{-1} to higher values such as 30 μg L^{-1} for industrial compounds in surface waters and up to 106 μg L^{-1} for the pharmaceutical ibuprofen in wastewaters. Compounds already banned in Europe such as atrazine, alkylphenols (APs) and alkylphenol polyethoxylates (APEOs) are still found in surface waters, nevertheless their origin is still poorly understood. Beyond the contamination of the Portuguese aquatic environment by PSs and EOPs, this review also highlights the need of more research on other classes of pollutants and emphasizes the importance of extending this research to other locations in Portugal, which have not been investigated yet.

The WFD (Directive 2000/60/EC) is one of the key legislation that delineates the requirements for monitoring water quality in the EU (EU-Directive 2000, 2013). This Directive promulgates an unified framework for Community action in the field of water policy to accomplish high water quality and adequate habitats for native flora and fauna, being applied to all waters bodies, e.g. lakes, rivers, groundwater bodies, transitional waters and coastal waters (EU-Directive 2000). Great efforts have been done in order to meet the overall quality status of the European waters. Recent reports of the WFD indicate that chemical quality of water bodies have significantly improved in the last 30 years. Nevertheless, a significant proportion of water bodies in EU will not accomplish the proposed objective of "good status" by the year 2015 (EU-Directive 2000, 2013). Evaluation of chemical quality of water bodies includes environmental impact assessment and monitoring data of a group of 45 PSs/Group of Substances that have been prioritized and updated for action at Community level (EU-Decision 2001; EU-Directive 2008, 2013). The PSs included in the Directive list are organic pollutants such as pesticides, hydrocarbons, organohalogenated compounds, APs, an organotin compound (tributyltin) and also toxic metals. Among these PSs, some of them have been classified as Priority Hazardous Substances because of their persistence and bioaccumulation rates, semivolatile properties and resistance to degradation (Ying et al. 2002; Vitali et al. 2004; Zhan et al. 2006; Wu et al. 2008; Zhou et al. 2008a, b; Ribeiro et al. 2015). Other organic pollutants such as hormones, industrial compounds and pharmaceuticals, recognized as EOPs have become major agents of concern for environmental studies due to their presence in water bodies and to their possible endocrine disrupting effects (Yokota et al. 2000; Waring and Harris 2005; Zuo et al. 2006; Xue et al. 2008; Wang et al. 2010; Wennmalm 2011; Wille et al. 2012). Thus, even detected at ng L^{-1} levels, their toxicity, persistence and bioaccumulation rates may produce adverse effects in both humans and wildlife (Werner et al. 2003; Piersma et al. 2009; Schwartz et al. 2010). In fact, substances such as diclofenac, 17-β-estradiol (E2) and 17-α-ethynylestradiol (EE2) were recommended for inclusion in the first watch list that shall establish ten compounds to be subjected to additional monitoring and toxicological data (EU-Directive 2013). Recently, the first watch list of substances to be monitored in the field of the water policy was launched and these substances are included together with estrone (E1), three macrolide antibiotics, pesticides and other organic compounds (EU-Decision 2015).

Portugal is located in southwestern Europe and has an extended coastal area. It is crossed by several rivers most of them flowing from east to west into the Atlantic Ocean. From north to south, several rivers run through such as the rivers Minho, Lima, Ave, Leça, Douro, Ria de Aveiro, Mondego, Tejo, Sado, Guadiana and Ria Formosa (see Fig. 1). Most of the Portuguese population is fixed in the coastal areas mainly in the north and center of the country, where most of industrial, agricultural and port activities are implemented. As a result, estuaries, rivers and coastal areas are subjected to an intensive anthropogenic pressure, by a wide range of pollutants discharged into them. Several works have addressed the ecological status of coastal waters and estuaries, mainly Douro, Mondego, Tejo and Sado (Mucha et al. 2004; Bordalo and Vieira 2005; Caeiro et al. 2005; Azevedo et al. 2006; Bordalo et al. 2006; Vasconcelos et al. 2007; Pereira et al. 2009; Echeandía et al. 2010). However, considering chemical status about specific pollutants, few publications refer to organic pollutants included in PSs list of the WFD or to EOPs in aquatic environment. Most works address the application of developed analytical methods and do not provide systematic monitoring data (spatio-temporal distribution) of Portuguese aquatic environment. These studies have demonstrated the presence of different classes of pollutants such as industrial compounds, natural and synthetic estrogens, phytoestrogens and phytosterols,

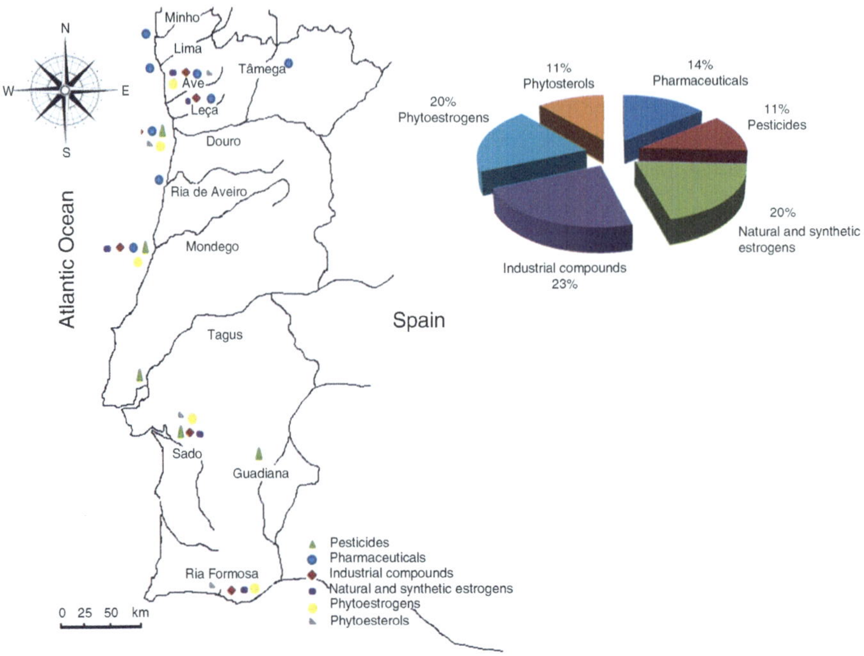

Fig. 1 Spatial distribution and frequency of occurrence reports of different classes of PSs and EOPs in Portuguese surface waters

pesticides, pharmaceuticals and personal care products in water compartments from Portugal (Quevauviller et al. 1989; Soares et al. 1999; Gil and Vale 2001; Ramalhosa et al. 2005; Rato et al. 2006; Silva et al. 2006; Gonçalves et al. 2007; Ribeiro et al. 2009a; Madureira et al. 2010; Nunes et al. 2011b). Hence, this paper intends to: a) summarize the occurrence of PSs included in the WFD list and some classes of EOPs in the aquatic environment of Portugal; b) identify those compounds that represent a high concern due to their concentration and/or toxicological effects; c) provide a comparative study of occurrence with other European countries; d) highlight classes of compounds that should be investigated. The search was based in ScienceDirect and ISI web of Knowledge databases considering articles published between 2001 and 2015 that comprise surface waters, ground and drinking waters and wastewaters as aquatic environmental matrices.

2 Surface Waters

This section describes the reports on occurrence of pollutants in surface waters such as: industrial compounds (e.g. bisphenol A (BPA), APs and APEOs), natural and synthetic estrogens (e.g. E1, E2 and EE2), phytoestrogens (e.g. daidzein, genistein, biochanin A), phytosterols (e.g. sitosterol), pharmaceuticals (e.g. antibiotics, cetirizine, carbamazepine, diazepam, fenofibric acid, propranolol, trimethoprim) and pesticides (e.g. organochlorine pesticides (OCPs), organophosphorous pesticides (OPPs), pyrethroids and triazines).

2.1 Industrial Compounds

Industrial compounds are the most studied compounds in surface waters (Azevedo et al. 2001; Quiros et al. 2005; Ribeiro et al. 2009a; Rocha et al. 2012a, 2013). These compounds include BPA mainly used in the manufacture of plastics and APs and APEOs widely used in several industrial and household products. Table 1 shows the locations sampled and concentration of these pollutants in various estuaries and rivers from Portugal. The presence of these compounds was first demonstrated by the studies of Azevedo et al. (2001) and Quiros et al. (2005). These authors measured concentrations of BPA, APs and APEOs in the order of μg L^{-1} in several surface waters from different origins of Portugal. In a monitoring study performed in 2005–2006 by Ribeiro et al. (2009a), BPA was found up to 10.7 μg L^{-1} in the Douro river estuary. This compound was found at lower concentrations in the Mondego estuary (up to 880.0 ng L^{-1}) (Ribeiro et al. 2009c) and in Sado estuary (up to 248.0 ng L^{-1}) (Ribeiro et al. 2009b). Recently, Rocha et al. (2011, 2012a, b, c, 2013a, b, c) evaluated the presence of BPA, several APs and APEOs included in the group of PSs of the WFD, in various estuaries and coastal areas (Rocha et al. 2012b, 2013a). These researchers reported the presence of BPA in Douro and Sado estuarine waters in 2009 and 2010 respectively, at lower concentrations than those found in the same estuaries by Ribeiro et al. (2009a, b). In the Ave estuary, considered one of the most polluted estuaries of Portugal, APEOs were the most abundant pollutants found in 2010 though they have been banned from Europe in 2003. These compounds were measured in all sampling points of the estuary and in the coastal areas. The frequency of these pollutants was 100 % and nonylphenol ethoxylates (NPEOs) were the most abundant compounds with concentration levels up to 750.6 ng L^{-1}. NPEOs were quantified at even higher concentrations in Douro estuary between 2009 and 2010, achieving 1147.5 ng L^{-1} (Rocha et al. 2013b). In Leça river, industrial compounds were also the most abundant pollutants and the NPEOs reached a maximum in seawater, namely 22.7 ng L^{-1} for NP1EO and 162.1 ng L^{-1} for NP2EO (Rocha et al. 2012c). Recent studies conducted in Sado river estuary showed the presence of nonylphenol up to 239.9 ng L^{-1} and NPEOs up to 1095.5 ng L^{-1}. The presence of these compounds was ubiquitous and even found in areas located in the Natural Reserve of this estuary (Rocha et al. 2012b). The origin of APs and APEOs was uncertain; nonetheless their occurrence was related to industrial activities and usage of sewage sludge as fertilizer. BPA was ubiquitously found in Ria Formosa at similar levels to those found in surface waters from other locations of Portugal. APEOs were quantified at higher levels in summer and at similar concentrations to other Portuguese surface waters (Rocha et al. 2013c).

Table 1 Occurrence of priority substances and emerging organic pollutants in Portuguese surface waters

River	Source	Sample collection	Class/Compound	Concentration	Reference
Minho	River water	March 2010–September 2011	Pharmaceuticals		Paíga et al. (2013)
			Ibuprofen	204.0 ng L^{-1}	
	River water	September-2011	Pharmaceuticals		Paíga et al. (2013)
			Paracetamol-glucuronide	0.18 µg L^{-1}	
			p-aminophenol	0.52 µg L^{-1}	
Lima	River water	March 2010–September 2011	Pharmaceuticals		Paíga et al. (2013)
			Ibuprofen	40.0–723.0 ng L^{-1}	
Tâmega	River water	September-2011	Pharmaceuticals		Paíga et al. (2013)
			p-aminophenol	0.40 µg L^{-1}	
Leça	River water		Estrogens		Rocha et al. (2012c)
			Estrone	0.5–7.4 ng L^{-1}	
			Estradiol	n.d.–13.4 ng L^{-1}	
			Ethynylestradiol	n.d.–2.8 ng L^{-1}	
			Industrial compounds		
			4-*t*-OP	5.2–8.3 ng L^{-1}	
			4-NP	n.d.–1.3 µg L^{-1}	
			4-OP	n.d.–5.4 ng L^{-1}	
			OP1EO	n.d.–2.9 ng L^{-1}	
			OP2EO	n.d.–131.0 ng L^{-1}	
			NP1EO	2.6–22.7 ng L^{-1}	
			NP2EO	38.9–162.1 ng L^{-1}	
	River water	September-2011	Pharmaceuticals		Paíga et al. (2013)
			Paracetamol-glucuronide	3.57 µg L^{-1}	
			p-aminophenol	1.25 µg L^{-1}	
			Paracetamol	0.25 µg L^{-1}	
	River water	March 2010–September 2011	Pharmaceuticals		Paíga et al. (2013)
			Ibuprofen	n.d.–256.0 ng L^{-1}	

Ave	River water and coastal line	January 2010–November 2010	Phytoestrogens		Rocha et al. 2013a
			Daidzein	8.6–74.3 ng L^{-1}	
			Genistein	52.4–682.3 ng L^{-1}	
			Biochanin A	213.5–398.1 ng L^{-1}	
			Formononetin	83.0–339.1 ng L^{-1}	
			Phytosterols		
			Sitosterol	671.8–5885.6 ng L^{-1}	
			Estrogens		
			Estrone	1.4–4.1 ng L^{-1}	
			Estradiol	1.6–5.9 ng L^{-1}	
			Ethynylestradiol	0.5–20.4 ng L^{-1}	
			Industrial compounds		
			BPA	7.9–521.8 ng L^{-1}	
			4-*t*-OP	3.4–25.4 ng L^{-1}	
			4-NP	59.0–154.9 ng L^{-1}	
			4-OP	0.6–4.4 ng L^{-1}	
			4-*n*-NP	0.8–16.8 ng L^{-1}	
			OP1EO	2.9–31.7 ng L^{-1}	
			NP1EO	48.6–227.8 ng L^{-1}	
			OP2EO	57.0–208.7 ng L^{-1}	
			NP2EO	199.5–750.6 ng L^{-1}	
	River water	September-2011	Pharmaceuticals		Paíga et al. (2013)
			Paracetamol-glucuronide	0.36 μg L^{-1}	
			p-aminophenol	1.63 μg L^{-1}	
			Paracetamol	0.17 μg L^{-1}	
	River water	March 2010–September 2011	Pharmaceuticals		Paíga et al. (2013)
			Ibuprofen	n.d–343 ng L^{-1}	

(continued)

Table 1 (continued)

River	Source	Sample collection	Class/Compound	Concentration	Reference
Douro	Estuarine water	December 2005–October 2006	Estrogens		Ribeiro et al. (2009a)
			Estrone	99.8–112.5 ng L^{-1}	
			Ethynylestradiol	56.0–101.9 ng L^{-1}	
			Phytoestrogens		
			Daidzein	14.6–888.4 ng L^{-1}	
			Genistein	29.1–197.4 ng L^{-1}	
			Biochanin A	17.1–191.4 ng L^{-1}	
			Industrial compounds		
			BPA	0.25–10.7 μg L^{-1}	
	Estuarine water	October 2007–July 2008	Pharmaceuticals		Madureira et al. (2010)
			Carbamazepine	0.37–178.0 ng L^{-1}	
			Diazepam	~3.65 ng L^{-1}	
			Fenofibric acid	1.48–70.3 ng L^{-1}	
			Propanolol	3.18 ng L^{-1}	
			Trimethoprim	3.89–15.7 ng L^{-1}	
			Sulfamethoxazole	9.14–53.3 ng L^{-1}	
	Estuarine water	March-2009	Industrial compounds		Rocha et al. (2011)
			4-t-OP	1.37–8.30 ng L^{-1}	
			4-NP	0.21–1.28 ng L^{-1}	
			NP1EO	5.90–22.70 ng L^{-1}	
			BPA	0.12–0.97 ng L^{-1}	
			OP1EO	0.73–5.71 ng L^{-1}	
			NP1EO	70.92–162.1 ng L^{-1}	
			Estrogens		
			Estrone	0.52–1.96 ng L^{-1}	
			Estradiol	6.25–14.36 ng L^{-1}	
			Ethynylestradiol	0.58–2.76 ng L^{-1}	

	Estuarine water	March 2009–May 2009	Pesticides		Rocha et al. 2012d
			Organochlorines	31.4–685.9 ng L^{-1}	
			Organophosphorous	23.2–467.8 ng L^{-1}	
			Pyrethroids and triazines	46.2–553.0 ng L^{-1}	
	River and its tributaries	March 2010–September 2011	Pharmaceuticals		Paíga et al. (2013)
			Ibuprofen	n.d–359 ng L^{-1}	
	Estuarine water	November 2009–September 2010	Industrial compounds		Rocha et al. 2013b
			BPA	20.4–313.7 ng L^{-1}	
			4-*t*-OP	5.1–30.6 ng L^{-1}	
			4-NP	88.2–170.0 ng L^{-1}	
			4-OP	<LOQ	
			4-*n*-NP	3.3–115.7 ng L^{-1}	
			OP1EO	33.0–60.4 ng L^{-1}	
			NP1EO	100.7–354.4 ng L^{-1}	
			OP2EO	100.4–424.0 ng L^{-1}	
			NP2EO	211.7–1147.5 ng L^{-1}	
			Phytoestrogens		
			Daidzein	6.7–24.2 ng L^{-1}	
			Genistein	16.6–137.8 ng L^{-1}	
			Biochanin A	728.4–19,091.1 ng L^{-1}	
			Formononetin	68.4–341.0 ng L^{-1}	
			Phytosterols		
			Sitosterol	8.7–2527.5 ng L^{-1}	
			Estrogens		
			Estrone	1.5–4.6 ng L^{-1}	
			Estradiol	5.4–8.5 ng L^{-1}	
			Ethynylestradiol	<LOQ–4.5 ng L^{-1}	

(continued)

Table 1 (continued)

River	Source	Sample collection	Class/Compound	Concentration	Reference
North and Central areas of Portugal	Surface water	not referred	Industrial compounds		Azevedo et al. (2001)
			BPA	up to 4.0 μg L^{-1}	
			NP	up to 30.0 μg L^{-1}	
Ria de Aveiro	Surface water	April to May 2010	Pharmaceuticals		Calisto et al. (2011)
			Carbamazepine	n.d.–0.11 μg L^{-1}	
			Cetirizine	n.d.–0.04 μg L^{-1}	
	Estuary and tributaries	March 2010–September 2011	Pharmaceuticals		Paíga et al. (2013)
			Ibuprofen	n.d.–229.0 ng L^{-1}	
Mondego	River water	April 2000–July 2000	Pesticides		Azevedo et al. (2004)
			Atrazine	up to 0.17 μg L^{-1}	
	River water	April 2000–July 2000	Pesticides	0.01–0.17 μg L^{-1}	Díez et al. (2005)
	Surface river water	November 2006–January 2007	Pharmaceuticals		Pena et al. (2007)
			Ciprofloxacin	79.6–119.2 ng L^{-1}	
			Enrofloxacin	67.0–102.5 ng L^{-1}	
	Estuarine water	December 2005–October 2006	Phytoestrogens		Ribeiro et al. (2009c)
			Daidzein	6.2–526.0 ng L^{-1}	
			Genistein	22.5–507.1 ng L^{-1}	
			Biochanin A	11.0–60.2 ng L^{-1}	
			Industrial compounds		
			BPA	106.4–880.0 ng L^{-1}	
Tejo	River basin	1983–1993	Pesticides		Cerejeira et al. (2003)
			Organochlorine and organophosphorous	2.0–1500 ng L^{-1}	
	River water	April 1999–May 2000	Organometalic compounds		Díez et al. (2005)
			TBT	20–25 ng Sn L^{-1}	

Sado	River water	April 2000–July 2000	Pesticides		Azevedo et al. (2001)
			Atrazine	up to 2.61 μg L^{-1}	
	River basin		Pesticides		Cerejeira et al. (2003)
		1998	Organochlorine and organophosphorous	0.068–48.0 μg L^{-1}	
		1999	Organochlorine and organophosphorous	0.012–31.6 μg L^{-1}	
	Estuarine water	July 2006 and December 2006	Phytoestrogens		Ribeiro et al. (2009b)
			Daidzein	160–500 ng L^{-1}	
			Genistein	100–320 ng L^{-1}	
			Biochanin A	10–150 ng L^{-1}	
			Industrial compounds		
			BPA	28.0–248 ng L^{-1}	
			Estrogens		
			Estrone	<LOQ	
			Estradiol	<LOQ	
			Ethynylestradiol	<LOQ	
	Estuarine water	February 2010–December 2010	Phytoestrogens		Rocha et al. (2012b)
			Daidzein	3.4–32.3 ng L^{-1}	
			Genistein	24.5–113.4 ng L^{-1}	
			Biochanin A	130.8–844.5 ng L^{-1}	
			Formononetin	423.4–2604.8 ng L^{-1}	
			Phytosterols		
			Sitosterol	63.9–4631.7 ng L^{-1}	
			Estrogens		
			Estrone	1.2–9.7 ng L^{-1}	
			Estradiol	7.2–10.8 ng L^{-1}	

(continued)

Table 1 (continued)

River	Source	Sample collection	Class/Compound	Concentration	Reference
			Ethynylestradiol	1.1–2.8 ng L^{-1}	
			Industrial compounds		
			BPA	12.2–28.9 ng L^{-1}	
			4-*t*-OP	11.4–22.0 ng L^{-1}	
			4-NP	129.2–239.9 ng L^{-1}	
			4-OP	2.8–26.0 ng L^{-1}	
			4-*n*-NP	2.6–27.3 ng L^{-1}	
			OP1EO	13.0–109.4 ng L^{-1}	
			NP1EO	60.0–311.4 ng L^{-1}	
			OP2EO	60.0–384.2 ng L^{-1}	
			NP2EO	167.0–1095.5 ng L^{-1}	
Ria Formosa	Surface water	February 2010–December 2010	Phytoestrogens		Rocha et al. (2013c)
			Daidzein	4.6–14.0 ng L^{-1}	
			Genistein	405.0–1157.0 ng L^{-1}	
			Biochanin A	91.0–261.0 ng L^{-1}	
			Formononetin	186.0–1041.0 ng L^{-1}	
			Phytosterol		
			Sitosterol	538.0–12,282.0 ng L^{-1}	
			Estrogens		
			Estrone	0.9–2.0 ng L^{-1}	
			Estradiol	4.7–10.1 ng L^{-1}	
			Ethynylestradiol	14.4–25.0 ng L^{-1}	
			Industrial compounds		
			BPA	6.5–72.0 ng L^{-1}	
			4-*t*-OP	4.3–41.0 ng L^{-1}	

			4-NP	12.2–547.0 ng L^{-1}	
			4-OP	0.5–8.5 ng L^{-1}	
			4-*n*-NP	3.4–14.6 ng L^{-1}	
			OP1EO	6.9–36.0 ng L^{-1}	
			NP1EO	41.0–279.0 ng L^{-1}	
			OP2EO	46.0–182.0 ng L^{-1}	
			NP2EO	49.0–780.0 ng L^{-1}	
Guadiana	River water	April 2000–July 2000	Pesticides		Azevedo et al. (2004)
			Atrazine	up to 0.79 μg L^{-1}	
	River basin	1993	Pesticides		Cerejeira et al. (2003)
			Organophosphorous and organo-chlorine pesticides	up to 0.3 μg L^{-1}	
Different origins in Portugal	Surface water	October 2005–February 2008	Pesticides		Carvalho et al. (2008)
			Atrazine	0.050–0.096 μg L^{-1}	
			Terbuthylazine	n.d.–0.132 μg L^{-1}	
			Desethylterbuthylazine	n.d.–0.056 μg L^{-1}	
Different origins in Portugal	River water		Industrial compounds		Quiros et al. (2005)
			BPA	up to 4.67 μg L^{-1}	
			OP	up to 30.23 μg L^{-1}	
			NP	up to 25.53 μg L^{-1}	
			NPEO1	up to 27.75 μg L^{-1}	
			NPEO2	up to 12.64 μg L^{-1}	

BPA: Bisphenol A; LOQ: limit of quantification; n.d.: not detected; 4-*n*-NP: 4-*n*-nonylphenol; NP1EO: nonylphenol monoethoxylate; NP2EO: nonylphenol diethoxylate; 4-*n*-OP: 4-*n*-octylphenol; 4-*t*-OP: 4-*t* octylphenol; OP1EO: 4-octylphenol monoethoxylate; OP2EO: octylphenol diethoxylate

Though various reports have demonstrated the presence of industrial compounds, their origin is still uncertain. Most of these studies were done in polluted rivers located near highly densely populated and industrialized areas. Thus, potential sources of industrial compounds in the surface waters were linked to discharges from domestic and industrial wastewaters. However, the occurrence of these pollutants in effluents from Wastewater Treatment Plants (WWTPs) was never investigated. Also, to date, studies concerning the occurrence of these pollutants in other environmental compartments, such as sediments that act as sink of these compounds, are not available. The seasonal occurrence is also diverse. Some reports demonstrated higher values during spring and autumn while others reported higher values during winter (Ribeiro et al. 2009a, b, c; Rocha et al. 2012a, b). This pattern demonstrates that sources of these pollutants can be diverse and related to random industrial activities or environmental conditions (such as precipitation, river flow, hydrological condition etc.), however further studies are needed to confirm these hypotheses.

2.2 *Natural and Synthetic Estrogens*

Natural and synthetic estrogens have been recognized as endocrine disrupting compounds (EDCs). Table 1 summarizes the concentrations of natural and synthetic estrogens found in aquatic matrices. These compounds were investigated in Ave, Leça, Douro and Sado rivers and also in Ria Formosa. Though estrogens can produce adverse effects in both humans and wildlife due to their toxicity, persistence and bioaccumulation rates even at ng L^{-1}, these compounds have not been investigated in other important Portuguese rivers such as Tejo or Guadiana. Ribeiro et al. (2009a, b) described a monitoring study in 2005–2006 of estrogens, in estuaries with different anthropogenic pressures. In the Douro river estuary, high concentrations of natural estrogens were observed, namely E1 up to 112.5 ng L^{-1} and the synthetic estrogen EE2 up to 101.9 ng L^{-1} (Ribeiro et al. 2009a). In the Sado river estuary, both E1 and EE2 were detected but not quantified, specifically in an industrial zone located near an urbanized area (Ribeiro et al. 2009b). Recently, Rocha et al. (2011, 2012a, b, c, 2013a, b, c) monitored all these estrogenic compounds in several estuaries (Table 1). E1, E2 and EE2 were found in the Leça river at levels up to 7.4 ng L^{-1}, 13.4 ng L^{-1} and 2.8 ng L^{-1}, respectively. Natural estrogens were quantified at lower levels in the Ave river, except EE2 which was found at higher values (up to 20.4 ng L^{-1}). During March of 2009, E1, E2 and EE2 attained levels in Douro river up to 1.96 ng L^{-1}, 14.36 ng L^{-1} and 2.76 ng L^{-1}, respectively. The concentrations reported in other study regarding the Douro river sampled between 2009 and 2010, were similar, with E1, E2 and EE2 reaching 4.6 ng L^{-1}, 8.5 ng L^{-1} and 4.5 ng L^{-1}, respectively. The concentrations found for E1 and EE2 in these reports are lower than those found earlier by Ribeiro et al. (2009a, b, c). The concentrations of E1, E2 and EE2 in Ria Formosa during 2010 were also in the order of few ng L^{-1} which is comparable to levels obtained in other rivers in the same period of time by Rocha et al. (2012a, b, c, 2013a, b).

The main inputs of natural and synthetic estrogens into the aquatic environment are human and animal excretion, through discharges from WWTPs and/or direct urban discharges. The investigated rivers and estuaries are located near densely populated areas and some of them are highly impacted by effluents from WWTPs. These compounds have been detected also in wastewaters (Sect. 4), corroborating WWTPs as a potential source of these pollutants in surface waters.

2.3 Phytoestrogens and Phytosterols

Phytoestrogens were evaluated for the first time in Portugal during a monitoring study comprising three estuaries, Douro, Mondego and Sado (Ribeiro et al. 2009a, b). These compounds are non-steroidal polyphenols with estrogenic activity although, lower than steroidal hormones such as E2. Phytoestrogens were quantified in high amounts, with daidzein being quantified in both Douro and Mondego estuaries, up to 888.4 ng L^{-1} and 526.0 ng L^{-1}, respectively (Ribeiro et al. 2009a, b). In the Sado river estuary phytoestrogens levels were lower compared to the other two estuaries, except for genistein which was inferior in Douro and biochanin A which was lower in Mondego (Ribeiro et al. 2009a, c). The overall lower concentrations in the Sado river are probably related with the high hydrodynamic of this estuary. These reports cited above were the first monitoring studies demonstrating the presence of these EDCs in Portuguese waters. Recently, Rocha et al. (2012b, c, 2013a, b, c) evaluated the presence of phytoestrogens and sitosterol (phytosterol) in several rivers. Also, phytoestrogens concentrations showed a seasonal fluctuation that may be related with the intense agriculture activity near the estuary. The higher values of daidzein and formononetin were observed during spring and summer while the highest concentrations for genistein and biochanin A were found during summer and autumn. Genistein showed the highest values up to 1157.0 ng L^{-1} and sitosterol up to 12,282.0 ng L^{-1} in Ria Formosa.

These reports demonstrated that the occurrence pattern and seasonality of phytoestrogens and phytosterols, as well as their potential sources in surface waters, seem to be highly diverse and are still poorly understood. Possible sources of these compounds have been suggested such as discharges from WWTPs and/or non-point sources, such as agricultural practices and/or lixiviation of local flora.

2.4 Pharmaceuticals

Awareness about the impact of pharmaceuticals in the environment has grown considerably. The scientific knowledge about the widespread distribution of these compounds in the aquatic environment and their potential harmful effects is a subject of major concern. Human and veterinary uses are the main cause of the presence of pharmaceuticals in the environment due to excretion and improper disposal of unused medicines. Various studies have been done to investigate the presence of these pollutants in wastewaters for evaluation of their contribution to the aquatic environmental contamination (Sect. 4).

The presence of antibiotics such as fluoroquinolones in surface waters was investigated by Pena et al. (2007), being found in samples collected at different points of the Mondego river near Coimbra, from November 2006 to January 2007. In this study, ciprofloxacin and enrofloxacin ranged from 79.6 to 119.2 ng L^{-1} and from 67.0 ng L^{-1} to 102.5 ng L^{-1}, respectively (Pena et al. 2007). The first systematic study regarding pharmaceuticals of several therapeutic classes in aquatic compartments was reported by Madureira et al. (2010) in a monitoring survey of Douro river estuary. This study allowed the determination of the distribution of six pharmaceuticals belonging to different therapeutic classes (Madureira et al. 2010). Carbamazepine (CBZ) was the most ubiquitous compound reaching levels up to 178 ng L^{-1}, which is consistent with the persistent pattern observed in other aquatic systems for this compound (Zhang et al. 2008). Maximum concentrations for all compounds were found at a sampling site located near the Sobreira WWTP. Concentrations found were up to 3.65 ng L^{-1} for diazepam, 70.3 ng L^{-1} for fenofibric acid, 3.18 ng L^{-1} for propranolol, 15.7 ng L^{-1} for trimethoprim and 53.3 ng L^{-1} for sulfamethoxazole. Beyond the presence of pharmaceuticals in the Douro river estuary, these results showed that their presence is consistent and constant during the time. In addition, this study showed the relation between the concentration of pharmaceuticals and the anthropogenic influence. Another study conducted by Calisto et al. (2011) demonstrated the presence of CBZ in surface waters from Ria de Aveiro (Calisto et al. 2011). This compound and cetirizine were quantified in 1 of the 11 sampling points of surface waters from Ria de Aveiro. Recently, other studies reported the determination of ibuprofen and its metabolites (Paíga et al. 2013) in several rivers in the north of Portugal, namely rivers Minho, Lima, Leça, Ave, Douro and Ria de Aveiro, being quantified up to 723 ng L^{-1}. The metabolite carboxy-ibuprofen has previously been detected in sediments of a tributary of the Douro river up to 46.1 μg kg^{-1} (Braganca et al. 2012). The presence of paracetamol, its principal metabolite (paracetamol-glucuronide) and its main transformation product (*p*-aminophenol) was investigated in various rivers located in the north of Portugal, such as Minho, Lima, Tâmega, Ave, Leça and Douro (Santos et al. 2013b). The parent drug was detected at lower concentrations than its metabolites, achieving 250 ng L^{-1} in the Leça river while metabolites were found at highest concentrations up to 1630 ng L^{-1} for *p*-aminofenol and up to 3570 ng L^{-1} for paracetamol-glucuronide in the Ave and Leça river, respectively. It is important to denote that to date all studies of occurrence of pharmaceuticals in Portuguese surface waters refer only to a single class of compounds, with the exception of one work dealing with determination of multi-therapeutic classes of pharmaceuticals (Madureira et al. 2010).

2.5 *Pesticides*

Pesticides are widely used in agricultural areas usually located near rivers and estuaries. Some of these compounds are lipophilic and trend to (bio)accumulate leading to an increase (Nunes et al. 2011a; Rodrigues et al. 2013) in estuarine and river sediments as well as bioconcentration in fat tissues of biota (Villaverde et al. 2008). Almost all classes of pesticides, OCPs such as dichlorodiphenyltrichloroethane (DDT), endosulfan, lindane, aldrin, dieldrin; OPPs such as dichlorvos, metolachlor, parathion; pyrethroids and triazines were found in estuarine waters (Table 1). The first monitoring study reported various OCPs and OPPs insecticides and herbicides in surface waters collected from 1983 to 1993, in three river basins of Tejo, Guadiana and Sado (Cerejeira et al. 2003). In this study, maximum values were found particularly for atrazine (up to 0.63 μg L^{-1}), chlorfenvinphos (up to 31.6 μg L^{-1}), α- and β-endosulfan (both up to 0.18 μg L^{-1}), lindane (up to 0.24 μg L^{-1}), molinate (up to 48.0 μg L^{-1}) and simazine (up to 0.3 μg L^{-1}). In the Tejo river basin, pesticides showed a seasonal variation with highest values observed during spring after pesticides application in the crops. Lindane, atrazine and molinate were the most often detected pesticides in the river basin of the Tejo near agricultural areas. In the upper zones of the river basin, pesticides levels were below detection levels corresponding to the forest areas. In the Guadiana river basin, pesticides also showed a seasonal pattern, with atrazine and simazine reaching the maximum values. In the Sado river basin, a seasonal variation was observed and molinate reached maximum values in June and July (48.0 μg L^{-1}) and chlorfenvinphos was quantified up to 31.6 μg L^{-1}. In another study reported by Azevedo et al. (2001) comprising Mondego, Sado and Guadiana rivers, it was observed that atrazine was the most common herbicide found at concentrations ranging from 0.79 to 2.61 μg L^{-1}. The highest values for atrazine were found in the Sado river during April sampling collection. In a study conducted by Rocha et al. (2012d), OCPs were detected in estuarine waters from the Douro river between 31.4 and 685.9 ng L^{-1}, OPPs between 23.2 and 467.8 ng L^{-1} and pyrethroids between 46.2 and 553.0 ng L^{-1}. Since pesticides are expected to be present in rivers and estuaries near agricultural areas and there is a lack of information regarding this kind of compounds in surface waters, it is of high importance to the scientific community to be aware of this fact.

2.6 *Portuguese* vs *European Surface Water: Occurrence of PSs and EOPs*

Many classes of environmental pollutants were reported in the Portuguese surface waters. The majority of the studies have been carried out in the north and to a lower extent in the center of Portugal. Only a limited number of research studies reported the presence of pollutants in the south. Figure 1 shows the spatial distribution of

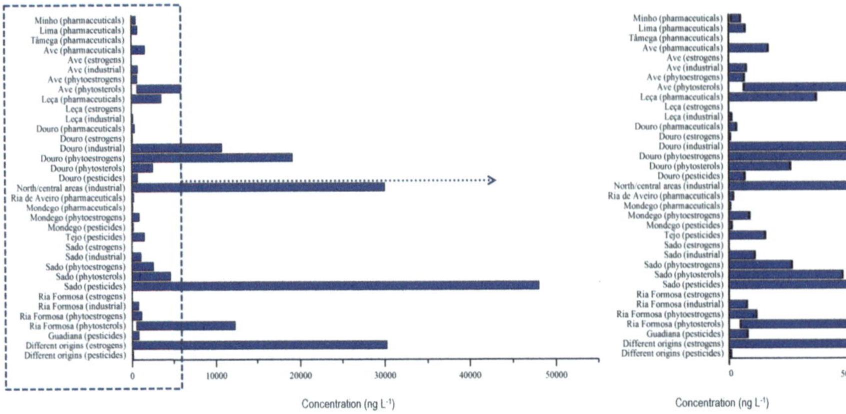

Fig. 2 Concentration range (ng L^{-1}) of the PSs and EOPs detected in the Portuguese surface waters

occurrence reports in Portuguese surface waters regarding different classes of pollutants and a graphical frequency of the different PSs and EOPs that have been found in surface waters in Portugal.

Regarding the levels of the detected organic pollutants, the range is very wide and varies highly depending on the compounds and the location (Fig. 2). The pesticide class was reported at the highest concentration in the Sado estuary and also in different origins of Portugal, followed by the class of industrial compounds in both north and central areas. Some estuaries such as Minho, Lima, Tâmega, Ria de Aveiro, Mondego, Tejo and Guadiana, were reported with lower levels of the micropollutants studied. This fact is probably due to the lack of studies on such areas, regarding many classes of substances.

Industrial and household compounds such as BPA, APs and APEOs were widely investigated in Portuguese surface waters. These pollutants were measured from few ng L^{-1} to levels up to 30 μg L^{-1} (APs) (Azevedo et al. 2001; Quiros et al. 2005; Ribeiro et al. 2009a; Rocha et al. 2013b). Data reported by Rocha et al. between 2009 and 2010, demonstrated that in some surface waters, concentrations of these pollutants are decreasing and levels measured are lower than the maximal allowed levels for surface waters described in the Directive 2013/39/EC (Rocha et al. 2011, 2012c, 2013a, b, c). These concentrations are within the range reported in surface waters in European countries such as Spain (up to 1.9 μg L^{-1}, APEOs), Switzerland (up to 69.0 μg L^{-1}, APEOs), Netherlands (up to 6.3 μg L^{-1}, APs), France (up to 0.43 μg L^{-1}, APs) (Bergé et al. 2012). EDCs comprising E1, E2 and EE2 were detected in Portuguese surface waters and were found always at ng L^{-1} levels (Ribeiro et al. 2009a; Rocha et al. 2011, 2013a, b, c). Though found at ng L^{-1} levels, these values are within the ranges reported to be able to promote endocrine disrupting effects in aquatic animals (Jobling et al. 1998, 2006). These compounds were found at levels higher than others reported in European countries such as Italy (up to 50 ng L^{-1}) (Meffe and Bustamante 2014). However, in the studies reported by Rocha et al. between 2009 and 2010, data showed comparable values to those found in surface waters from rivers and coastal areas of Belgium, Netherlands and France (Belfroid et al. 1999; Baronti et al. 2000; Cargouet et al. 2004, 2007; Noppe et al. 2007). E2, E1 and EE2 were recently included in the first watch list of EU-Decision (2015), in order to gather monitoring data for the purpose of facilitating the determination of appropriate measures to address the risk posed by those substances (EU-Directive 2013). Phytoestrogens are well described in most Portuguese surface waters though they have not been investigated in coastal waters and in the Tagus estuary, the largest of Portugal. In those studies, daidzein, genistein, biochanin A, formononetin and the phytosterol sitosterol, were always detected at μg L^{-1} range (Rocha et al. 2013). Biochanin A was found in the Douro river at approximately 19 μg L^{-1} (Rocha et al. 2013b); formononetin was quantified over 1 μg L^{-1} in Sado river (Rocha et al. 2012b); and both formononetin and genistein were detected above 1 μg L^{-1} in Ria Formosa (Rocha et al. 2013c). The other phytoestrogens were detected at ng L^{-1} levels (Ribeiro et al. 2009a, b). These data were similar to others reported for surface waters in Italy (Lagana et al. 2004; Bacaloni et al. 2005) and Switzerland (Erbs et al. 2007). Pharmaceuticals in surface waters need more research to keep up with the studies performed in wastewaters matrices, as will be discussed in Sect. 4.1. In fact, there are only some publications in the north of Portugal that cover some therapeutic classes and most of them are related to application of developed methods and not to monitoring data (Pena et al. 2007; Calisto et al. 2011; Paíga et al. 2013) with exception of the work developed by Madureira et al. (2010). In the south of Portugal there is not available data about the presence of pharmaceuticals in surface waters. This means that there is a lack about the occurrence of these EOPs in surface waters from Portugal. The few available data demonstrated that these compounds are present in Portuguese surface waters. The highest concentration was found for paracetamol and its metabolites and CBZ was the most frequently detected drug. Nevertheless, various studies have been done in other European countries (da Silva et al. 2011; Gros et al. 2012; Osorio et al. 2012; Carmona et al. 2014; Esteban et al. 2014) and all around the world (Boyd et al. 2003; Lin and Tsai 2009; Yoon et al. 2010; Blair et al. 2013; Liu and Wong 2013; Writer et al. 2013). Various monitoring data have demonstrated the seasonality of these pollutants and their negative effects in the aquatic environment.

Pesticide compounds, namely OCPs, OPPs, pyrethroids and triazines classes have been found in Portuguese waters in the range of ng to μg L^{-1} (Azevedo et al. 2001, 2004; Cerejeira et al. 2003; Díez et al. 2005; Carvalho et al. 2008; Rocha et al. 2012c). The PSs atrazine, simazine and DDT were found in surface waters above the admissible environmental levels (EU-Directive 2008). Also, some of these studies were occasional collections or application of developed methods rather than planned monitoring studies. This fact suggests that more studies shall be done urgently in order to have a global evaluation about the presence of pesticides included in the PSs list, as other classes of these compounds in Portugal.

3 Groundwater and Drinking Waters

3.1 Pesticides

Drinking water for human consumption during the production (raw water) and distribution phases (treated water) has been checked for the presence of pesticides (Table 2). In the study developed by Carvalho et al. (2008) during the period 2005–2008, 1600 water samples used for human consumption and ground and surface waters were collected in different parts of Portugal to verify the presence of the pesticides atrazine, desethylatrazine, terbuthylazine, desethylterbuthylazine and diuron. Concentration levels of pesticides in drinking water ranged from 0.031 to 0.259 μg L^{-1} (Carvalho et al. 2008). The concentrations found in groundwaters were in the same range as those obtained for drinking waters (few μg L^{-1}). Groundwater samples collected from Póvoa de Varzim, an area of intensive agriculture in Portugal, demonstrated the presence of other pesticides in the same level of concentration as those found in the study of Carvalho et al. The quantifiable values of chlorfenvinphos ranged from 0.2 to 0.303 μg L^{-1} and endosulfan from 0.007 to 0.117 μg L^{-1} (Farré et al. 2002). Since groundwater in this area of Portugal in not only used for agricultural irrigation purposes but also for human consumption, this fact is of major importance. Groundwater samples collected in the Mondego river drainage area located in a strongly anthropogenic and agricultural area, showed pesticides concentration levels above regulatory levels of 0.5 μg L^{-1} in 32 % of the analyzed samples (Andrade and Stigter 2009).

3.2 Pharmaceuticals

The first systematic monitoring program concerning the occurrence of pharmaceuticals in a water supply system for production of drinking water was investigated in a study developed by Gaffney et al. (2015) in partnership with EPAL (Empresa Portuguesa das Águas Livres S.A.) in the Lisbon area. In this study, several classes of pharmaceuticals were investigated in the water supply system (groundwater and surface water) and drinking water. In addition, human health risk assessment was also evaluated based on risk quotients. From the 31 investigated compounds, 16 were found in the water system supply. The pharmaceuticals concentrations ranged from 0.03 to 46 ng L^{-1}, from 0.11 to 30 ng L^{-1} and from 0.005 ng L^{-1} to 12 ng L^{-1} in the River Tagus, River Zêzere and groundwater, respectively. The Tagus River presented the highest values of pharmaceuticals which is in agreement with the high impact of WWTPs. Most of the pharmaceuticals were not found in the drinking water. Only eight compounds were removed during the drinking water treatment processes, being caffeine, carbamazepine, atenolol, sulfadiazine, sulfapyridine, sulfamethoxazole and erythromycin not removed and still quantified in drinking water.

Table 2 Occurrence of priority substances and emerging organic pollutants in Portuguese groundwaters and drinking waters

Origin	Source	Sample collection	Compound	Concentration	Reference
Póvoa de Varzim	Groundwater	Not referred	Pesticides		Farré et al. (2002)
			Ametryn	<LOQ	
			Chlorfenvinphos	n.d.–0.303 μg L^{-1}	
			Endosulfan	n.d.–0.117 μg L^{-1}	
Aveiro	Groundwater (geographical area of Ria de Aveiro)	April to May 2010	Pharmaceuticals		Calisto et al. (2011)
			Carbamazepine	n.d.	
			Cetirizine	n.d.	
Mondego river drainage basin	Groundwater	March to July, 2001 and 2002, October 2001 and September 2002	Pesticides		Andrade and Stigter (2009)
			Alachlor	n.d.–0.83 μg L^{-1}	
			Atrazine	n.d.–0.49 μg L^{-1}	
			Metolachlor	n.d.–0.06 μg L^{-1}	
			Molinate	n.d.–3.62 μg L^{-1}	
			Propanil	n.d.–2.07 μg L^{-1}	
			3,4-dichloroaniline	n.d.–13.36 μg L^{-1}	
Baixo Sado	Groundwater	2002–2003	Pesticides		Silva et al. (2006)
			Molinate	up to 59.46 μg L^{-1}	
			Propanil	up to 1.86 μg L^{-1}	
			3,4-dichloroaniline	up to 0.71 μg L^{-1}	
			Endosulfan	up to 0.2 μg L^{-1}	
			Oxadiazon	up to 0.13 μg L^{-1}	
			MCPA	up to 0.09 μg L^{-1}	
			Profoxydim	up to 0.07 μg L^{-1}	
			Cycloxydim	up to 0.03 μg L^{-1}	
Different origins in Portugal	Groundwater	October 2005–February 2008	Pesticides		Carvalho et al. (2008)
			Atrazine	0.052–0.062 μg L^{-1}	
			Terbuthylazine	n.d.–0.027 μg L^{-1}	
			Desethylterbuthylazine	n.d.–0.094 μg L^{-1}	

(continued)

Table 2 (continued)

Origin	Source	Sample collection	Compound	Concentration	Reference
Different origins in Portugal	Drinking water	October 2005–February 2008	Pesticides		Carvalho et al. (2008)
			Atrazine	0.046–0.162 μg L^{-1}	
			Terbuthylazine	0.031–0.130 μg L^{-1}	
			Diuron	0.036–0.259 μg L^{-1}	
Lisbon	EPAL water supply system (surface water from Tagus, Zêzere and groundwater)	January 2013–December 2013	Pharmaceuticals		Gaffney et al. (2015)
			Caffeine	up to ~46 ng L^{-1}	
			Indomethacin	up to ~36 ng L^{-1}	
			Erythromycin	up to ~31 ng L^{-1}	
			Acetaminophen	up to ~31 ng L^{-1}	
			Nimesulide	up to ~28 ng L^{-1}	
			Sulfadiazine	up to ~27 ng L^{-1}	
			Sulfamethoxazole	up to ~23 ng L^{-1}	
			Ibuprofen	up to ~21 ng L^{-1}	
			Gemfibrozil	up to ~16 ng L^{-1}	
			Carbamazepine	up to ~18 ng L^{-1}	
			Diclofenac	up to ~12 ng L^{-1}	
			Naproxen	up to ~6 ng L^{-1}	
			Sulfapyridine	up to ~8 ng L^{-1}	
			Atenolol	up to ~4 ng L^{-1}	
			Propranolol	up to ~1 ng L^{-1}	
			Sulfamethazine	up to ~1 ng L^{-1}	
Lisbon	Drinking water	2011–2012	Total pesticides	<0.1 μg L^{-1}	EPAL (2012)
			Polycyclic Aromatic Hydrocarbon	<0.1 μg L^{-1}	

LOQ: limit of quantification; n.d.: not detected

3.3 Portuguese vs *European Groundwater and Drinking Water: Occurrence of PSs and EOPs*

To date, studies in Portugal concerning the occurrence of PSs and EOPs such as industrial compounds, natural and synthetic estrogens among others, in groundwater and drinking water are not available. These environmental matrices are so far the least studied. This may be due to the higher difficulty of groundwater sampling and generally lower concentrations found either in ground and drinking water, comparatively to surface waters. Nevertheless, there are few studies in Portugal that confirm the presence of pesticides and there is only one study about the occurrence of pharmaceuticals in groundwater and even in drinking water (Farré et al. 2002; Carvalho et al. 2008; Andrade and Stigter 2009, Gaffney et al. 2015). In fact, diuron exceeded the threshold in 9 of the 13 drinking water samples. These results revealed the difficult of eradicating these compounds, even after water treatment, and the hazards to human health. The PS atrazine was the most frequently detected pesticide in groundwater, even after being banned from EU. Thus, investigation in this area should be immediately implemented. Comparing to other European countries, pesticides values were within values found in Spain (Jurado et al. 2012; Postigo and Barceló 2015), Italy (Meffe and Bustamante 2014) and France (Morvan et al. 2006; Baran et al. 2007). Nevertheless, the incomplete removal during the WWTPs treatment processes and transformation products that can be formed during this process or in the environment, as well as the potential adverse effects from exposure to mixtures of various classes of pollutants, their metabolites and related transformation products, is still poorly understood and it is crucial to the assessment of human health risks (Kim et al. 2007; Schriks et al. 2010).

4 Influent and Effluent Wastewaters

4.1 Pharmaceuticals

Table 3 synthesizes the most important studies performed in Portugal about pharmaceuticals in wastewaters. Various therapeutic classes have been found in wastewater effluents, demonstrating the difficulty to remove these compounds, even after wastewater treatment. This fact highlights the risk for human health and the environmental impact of these compounds in non-targeted organisms. The antiepileptic drug CBZ is among the most frequently detected pharmaceutical, being indicated as anthropogenic pollution marker (Salgado et al. 2010; Calisto et al. 2011; Sousa et al. 2012). CBZ and cetirizine were quantified in all wastewaters samples at ng L^{-1} level in a study developed by Calisto et al. 2011 conducted in

Table 3 Occurrence of priority substances and emerging organic pollutants in Portuguese influents and effluents from WWTPs

Location	Source	Sample collection	Compound	Concentration	Reference
North of Portugal	Wastewater	December 2008–March 2009	Polycyclic musk fragrances		Machado et al. (2011)
			Isoborneol	n.d.–471 ng L^{-1a}	
			Galaxolide	222–444 ng L^{-1a}	
Porto	WWTP effluents located along Douro extension	March 2010–September 2011	Pharmaceuticals		Paíga et al. (2013)
	Hospital effluents of the metropolitan area of Porto		Ibuprofen	n.d.–616 ng L^{-1}	
				1802–3868 ng L^{-1}	
	Landfill leachate			9761–48,720 ng L^{-1}	
	WWTPs discharging to Douro river (influents and effluents)	2009 (every 3 months)	Pharmaceuticals		Sousa et al. (2011)
			Paracetamol	n.d.–20,074 ng L^{-1}	
			Diclofenac	431–1597 ng L^{-1}	
			Ibuprofen	1088–3588 ng L^{-1}	
			Ketoprofen	122–562 ng L^{-1}	
			Naproxen	161–3474 ng L^{-1}	
			Bezafibrate	343–5217 ng L^{-1}	
			Gemfibrozil	290–1183 ng L^{-1}	
			Simvastatin	n.d.–1255 ng L^{-1}	
			Azithromycin	530–836 ng L^{-1}	
			Lorazepam	n.d.–438 ng L^{-1}	
			Paroxetine	45–240 ng L^{-1}	
			Furosemide	n.d.–2675 ng L^{-1}	
			Hydrochlorothiazide	n.d.–6022 ng L^{-1}	
			Bisoprolol	187–397 ng L^{-1}	

	Febros WWTP, North of Portugal (effluent)	March 2011	Pharmaceuticals		Sousa et al. (2012)
			Diclofenac	24,256 ng L^{-1}	
			Ketoprofen	410 ng L^{-1}	
			Fenofibrate	53 ng L^{-1}	
			Gemfibrozil	215 ng L^{-1}	
			Azithromycin	631 ng L^{-1}	
			Ciprofloxacin	254 ng L^{-1}	
			Ofloxacin	101 ng L^{-1}	
			Norfloxacin	138 ng L^{-1}	
			Alprazolam	244 ng L^{-1}	
			Lorazepam	682 ng L^{-1}	
			Fluoxetine	24 ng L^{-1}	
			Paroxetine	29 ng L^{-1}	
			Furosemide	492 ng L^{-1}	
			Hydrochlorothiazide	3051 ng L^{-1}	
			Bisoprolol	132 ng L^{-1}	
			Propanolol	52 ng L^{-1}	
			Carvedilol	95 ng L^{-1}	
			Losartan	149 ng L^{-1}	
			Clotrimazole	12 ng L^{-1}	
			Fluconazole	110 ng L^{-1}	
			Terbinafine	37 ng L^{-1}	
			Carbamazepine	417 ng L^{-1}	

(continued)

Table 3 (continued)

Location	Source	Sample collection	Compound	Concentration	Reference
	Febros WWTP, North of Portugal (influent)	Two consecutive 48 h periods over two successive weeks	Pharmaceuticals		Salgado et al. (2011)
			Atenolol	n.d.–4341 ng L^{-1}	
			Clorazepate	n.d.–3332 ng L^{-1}	
			Etofenamate	n.d.–40,168 ng L^{-1}	
			Fluoxetine	n.d.–3465 ng L^{-1}	
			Paroxetine	n.d.–39,732 ng L^{-1}	
			Hydroxyzine	n.d.–1168 ng L^{-1}	
			Indapamide	n.d.–15,386 ng L^{-1}	
			Captopril	n.d.–4321 ng L^{-1}	
			Clofibric acid	n.d.–41,428 ng L^{-1}	
			Diclofenac	n.d.–64,479 ng L^{-1}	
			Enalapril	n.d.–6244 ng L^{-1}	
			Ibuprofen	n.d.–52,201 ng L^{-1}	
			Ketoprofen	up to 104,114 ng L^{-1}	
			Ampicillin	n.d.–252 ng L^{-1}	
			Estrone	n.d.–177 ng L^{-1}	
	3 WWTPs from Portugal (effluent)	October-2013	Pharmaceuticals		Ribeiro et al. (2014)
			Fluoxetine	n.d.–<LOQ	
			Norfluoxetine	n.d.	
			Venlafaxine	40.4–129 ng L^{-1}	
			Salbutamol	n.d.	
			Alprenolol	n.d.	
			Bisoprolol	<LOQ	
			Metoprolol	n.d.–<LOQ	
			Propranolol	<LOQ	

	Febros WWTP, North of Portugal (effluent)	March-2011	Polycyclic musk fragrances		Sousa et al. (2012)
			2,4,6-Trichloroanisole	226 ng L^{-1}	
			Galaxolide	3604 ng L^{-1}	
			Musk ketone	134 ng L^{-1}	
			Tonalide	703 ng L^{-1}	
	Febros WWTP, North of Portugal (influent)	Two consecutive 48 h periods over two successive weeks	Polycyclic musk fragrances		Salgado et al. (2011)
			Galaxolide	50–2780 ng L^{-1}	
			Tonalide	18–816 ng L^{-1}	
			Cashmeran	66–4040 ng L^{-1}	
			Celestolide	6–1442 ng L^{-1}	
			Traseolide	7–676 ng L^{-1}	
	Porto (effluent wastewater from a textile industry)	June-1999	Polyethoxylate surfactants		Farré et al. (2001)
			polyethylene glycol 2	1210 μg L^{-1}	
			C10EO6 polyethoxylated decanol	n.d.–3513 μg L^{-1}	
			Bis (2-ethylhexyl) phthalate	n.d.–173 μg L^{-1}	
			3-nitro-BS	n.d.–132 μg L^{-1}	
			2-NPS	n.d.–2377 μg L^{-1}	
			1-Hidroxy-4-NPS	n.d.–21.4 μg L^{-1}	
Aveiro	Wastewater (geographical area of Ria de Aveiro)	April to May 2010	Pharmaceuticals		Calisto et al. (2011)
			Carbamazepine	n.d.–0.65 μg L^{-1}	
			Cetirizine	n.d.–0.60 μg L^{-1}	

(continued)

Table 3 (continued)

Location	Source	Sample collection	Compound	Concentration	Reference
Coimbra	Wastewater samples from four hospitals of Coimbra (influent and effluent)	Spring and Autumn 2007	Pharmaceuticals		Seifrtová et al. (2008)
			Ciprofloxacin	100.8–10,962.5 ng L^{-1}	
			Ofloxacin	up to 10,675.5 ng L^{-1} (only in hospitals)	
			Norfloxacin	up to 455.0 ng L^{-1}	
			Enrofloxacin	up to 447.1 ng L^{-1} (only in WWTP)	
	Effluent wastewaters from 4 hospitals of Coimbra (university, general, pediatric and maternity hospitals) and a WWTP receiving domestic and the four hospital effluents	February–May 2011 (9 sampling periods for hospitals and 7 for influents and effluents of WWTP)	Pharmaceuticals		Santos et al. (2013a)
			14 analgesics and anti-inflammatories	up to 58,857 ng L^{-1}	
			5 lipid regulators and cholesterol lowering statin drugs	up to 2086 ng L^{-1}	
			13 psychiatric drugs	up to 2042 ng L^{-1}	
			2 histamine H_1receptors antagonists	up to 10.2 ng L^{-1}	
			3 histamine H_2receptors antagonists	up to 19,840 ng L^{-1}	
			6 beta-blockers	up to 8037 ng L^{-1}	
			3 diuretics	up to 32,558 ng L^{-1}	
			2 oral antidiabetics	up to 4040 ng L^{-1}	
			4 antihypertensives	up to 19,822 ng L^{-1}	
			1 antiplatelet agent	up to 396 ng L^{-1}	
			1 prostatic hyperplasia	up to 3.20 ng L^{-1}	
			1 β-agonist	up to 2595 ng L^{-1}	

			1 anticoagulant	up to 8.28 ng L^{-1}	
			1 X-ray contrast	up to 611,429 ng L^{-1}	
			3 antihelmintics	up to 1746 ng L^{-1}	
			1 synthetic glucocorticoid	up to 352 ng L^{-1}	
			1 sedation and muscle relaxation	up to 24.4 ng L^{-1}	
			2 tranquilizer	up to 3.87 ng L^{-1}	
			11 antibiotics	up to 38,689 ng L^{-1}	
			3 calcium channel blockers	up to 1470 ng L^{-1}	
Seixal	WWTP of Fernão Ferro, Seixal, Portugal (influent, effluent and sludge)	Two successive weeks	Pharmaceuticals		Salgado et al. (2012)
			Atenolol	up to 12.80 g day^{-1}	
			Clorazepate	up to 1.83 g day^{-1}	
			Etofenamate	up to 25.33 g day^{-1}	
			Fluoxetine	up to 3.67 g day^{-1}	
			Hydroxyzine	up to 7.38 g day^{-1}	
			Indapamide	up to 8.64 g day^{-1}	
			Captopril	up to 2.88 g day^{-1}	
			Enalapril	up to 3.81 g day^{-1}	
			Clofibric acid	up to 49.88 g day^{-1}	
			Diclofenac	up to 131.61 g day^{-1}	
			Ibuprofen	up to 46.00 g day^{-1}	
			Ketoprofen	up to 83.34 g day^{-1}	
			Ampicillin	up to 0.52 g day^{-1}	
			Paroxetine	up to 32.93 g day^{-1}	

(continued)

Table 3 (continued)

Location	Source	Sample collection	Compound	Concentration	Reference
	WWTP of Fernão Ferro, Seixal, Portugal (influent, effluent and sludge)[a]	Two successive weeks	Polycyclic musk fragrances		Salgado et al. (2012)
			Galaxolide	up to 1.41 g day^{-1}	
			Tonalide	up to 0.61 g day^{-1}	
			Cashmeran	up to 4.63 g day^{-1}	
			Celestolide	up to 1.60 g day^{-1}	
			Traseolide	up to 0.71 g day^{-1}	
Lisbon and Tagus Valley	5 WWTP from Portugal (influent, effluent and sludge[a])	Spring (23 May–7 July) and autumn (2–25 October)	Pharmaceuticals		Salgado et al. (2010)
			Atenolol	n.d.–4757 ng L^{-1}/n.d.	
			Caffeine	n.d.–36,160 ng L^{-1}/ 1788–8423 ng g^{-1}	
			Carbamazepine	n.d.–994 ng L^{-1}/n.d.	
			Clorazepate	n.d.–6227 ng L^{-1}/ 181 ng g^{-1}	
			Dimethylaminophenazone	n.d.–4278 ng L^{-1}/ 158–1361 ng g^{-1}	
			Domperidone	n.d.–163 ng L^{-1}/n.d.	
			Etofenamate	n.d.–7333 ng L^{-1}/ 24,785–134,431 ng g^{-1}	
			Fentiazac	n.d.–5297 ng L^{-1}/n.d.	
			Fluoxetine	n.d.–1704 ng L^{-1}/ 77 ng g^{-1}	
			Fluticasone	n.d.–2848 ng L^{-1}/ 1473–2330 ng g^{-1}	
			Hydroxazine	n.d.–9344 ng L^{-1}/ 43,339 ng g^{-1}	

			Indapamide	n.d.–1236 ng L^{-1}/47–1362 ng g^{-1}	
			Nimesulide	n.d.–6911 ng L^{-1}/n.d.	
			Paroxetine	n.d.–3367 ng L^{-1}/n.d.	
			Piroxicam	n.d.–9298 ng L^{-1}/n.d.	
			Ramipril	n.d.–5445 ng L^{-1}/488 ng g^{-1}	
			Salbutamol	n.d.–2158 ng L^{-1}/12–104 ng g^{-1}	
			Tramadol	n.d.–1344 ng L^{-1}/n.d.	
			Captopril	n.d.–13,335 ng L^{-1}/875–5516 ng g^{-1}	
			Clofibric acid	n.d.–7286 ng L^{-1}/117–15,655 ng g^{-1}	
			Diclofenac	n.d.–6674 ng L^{-1}/2259–17,785 ng g^{-1}	
			Enalapril	n.d.–19,888 ng L^{-1}/61 ng g^{-1}	
			Flurbiprofen	n.d.–9631 ng L^{-1}/1018–3544 ng g^{-1}	
			Furosemide	n.d.–15,244 ng L^{-1}/3602 ng g^{-1}	
			Ibuprofen	n.d.–106,490 ng L^{-1}/550–3398 ng g^{-1}	
			Indomethacin	n.d.–8899 ng L^{-1}/20–88 ng g^{-1}	
			Ketoprofen	n.d.–14,275 ng L^{-1}/47–21,989 ng g^{-1}	
			Naproxen	n.d.–2894 ng L^{-1}/n.d.	

(continued)

Table 3 (continued)

Location	Source	Sample collection	Compound	Concentration	Reference
			Amoxicillin	n.d.–5698 ng L^{-1}/112–166 ng g^{-1}	
			Ampicillin	n.d.–4120 ng L^{-1}/n.d.	
			Estrogens		
			17-α-ethynylestradiol	n.d.–106 ng L^{-1}/221 ng g^{-1}	
			Estrone	n.d.–2484 ng L^{-1}/8–181 ng g^{-1}	
			β-estradiol	n.d.–344 ng L^{-1}/n.d.	
5 regions (North, Center, Lisbon and Tagus Valley, Alentejo and Algarve)	15 WWTPs (influents and effluents)	2013 (spring and summer)	Pharmaceuticals		Pereira et al. (2015)
			Alprazolam	n.d.	
			Lorazepam	up to 475.8 ng L^{-1}	
			Zolpidem	n.d.	
			Azithromycin	up to 719.3 ng L^{-1}	
			Ciprofloxacin	up to 17,500 ng L^{-1}	
			Bezafibrate	up to 6000 ng L^{-1}	
			Gemfibrozil	up to 4300 ng L^{-1}	
			Simvastatin	up to 8500 ng L^{-1}	
			Diclofenac	up to 2400 ng L^{-1}	
			Ibuprofen	up to 8600 ng L^{-1}	
			Paracetamol	up to 66,700 ng L^{-1}	
	15 WWTPs (influents and effluents)	2013 (four seasons)	Pharmaceuticals		Silva et al. (2014)
			Citalopram	up to 213.60 ng L^{-1}	
			Fluoxetine	up to 157.40 ng L^{-1}	
			Paroxetine	up to 186.40 ng L^{-1}	
			Sertraline	up to 1.67 ng L^{-1}	

LOQ: limit of quantification; n.d.: not detected; WWTP: wastewater treatment plant

[a]Loads are given, when the concentrations are not available

Aveiro area (Calisto et al. 2011). The presence of antibiotics has been also widely investigated. Fluoroquinolones were detected in wastewater samples from the four hospitals of Coimbra, as well as, in influents and effluents of a municipal WWTP in the same city (Seifrtová et al. 2008). In this study, ciprofloxacin, ofloxacin, norfloxacin enrofloxacin were found up to 10,962.5 ng L^{-1}, 10,675.5 ng L^{-1}, 455.0 ng L^{-1} and 447.1 ng L^{-1}, respectively. Their removals rates were high even in the autumn, which was related to the warm weather during that season: 85 % and 92 % for norfloxacin, 54 % and 76 % for ciprofloxacin, and 53 % and 56 % for enrofloxacin, in spring and autumn, respectively. Salgado et al. (2010) reported the presence of pharmaceuticals in influent, effluent and sludge of WWTPs during two seasons. These researchers monitored 65 pharmaceuticals in 5 WWTPs of Portugal and found their presence in the influent, effluent and secondary sludge between few ng L^{-1} and μg L^{-1}, most frequently in the influent. The nonsteroidal anti-inflammatory drugs (NSAIDs) (namely diclofenac, ibuprofen and flurbiprofen), the antihypertensives (namely captopril and enalapril), caffeine and clofibric acid were found at high concentrations in both influent and effluent of WWTPs, suggesting a poor removal rate of such compounds occurring in the WWTPs. The same group (Salgado et al. 2011) assessed the variability of pharmaceuticals in a full-scale activated sludge plant included in the previous work and found also high frequency and concentrations of some classes of pharmaceuticals (namely NSAIDs, antihypertensives, antidepressants, atenolol and clofibric acid). A recent determination of 23 pharmaceuticals in influents and effluents of two WWTPs of Porto was performed by Sousa et al. (2011). This study demonstrated the presence of various classes of pharmaceuticals in decreasing order of concentrations, paracetamol ranging from not detected to 20 μg L^{-1}, followed by hydrochlorothiazide, bezafibrate, ibuprofen, naproxen, furosemide, diclofenac, simvastatin and gemfibrozil. Other pharmaceuticals such as azithromycin, ketoprofen, lorazepam, bisoprolol and paroxetine were quantified in decreasing order at levels below 1 μg L^{-1}. All these pharmaceuticals were poorly removed in the two studied WWTPs, being noteworthy that in many cases, the concentrations quantified in the effluents were higher than those found in the influent samples, probably due to spot sampling or hydrolysis of conjugated metabolites during the treatment processes, originating the parent substances. The same group quantified 22 pharmaceuticals in the effluent of Febros WWTP and many pharmaceuticals were detected, with a high concentration for diclofenac (24,256 ng L^{-1}), which was apparently poorly removed, followed by hydrochlorothiazide (3051 ng L^{-1}) (Sousa et al. 2012). Other pharmaceuticals were quantified below 1 μg L^{-1} such as lorazepam, azithromycin, furosemide, carbamazepine and ketoprofen. Pharmaceuticals quantified below 500 ng L^{-1} included fluoroquinolones, antidepressants, beta-blockers and antifungals. Recently, other studies reported the determination of ibuprofen and its metabolites (Paíga et al. 2013) in several WWTP effluents up to 48,720 ng L^{-1}. Santos et al. (2013a) monitored the occurrence of 78 pharmaceuticals belonging to several therapeutic classes in hospital effluents and influents as well as effluents of municipal WWTPs, with remarkable concentrations namely up to 611,429 ng L^{-1} for the X-ray contrast iopromide, up to

58,857 ng L^{-1} for the analgesics and anti-inflammatories class, up to 38,689 ng L^{-1} for the antibiotics class, up to 32,558 ng L^{-1} for the diuretics class, up to 19,822 ng L^{-1} for the antihypertensives class, up to 8037 ng L^{-1} for the beta-blockers class, as examples of the highest concentrations. Four therapeutic classes (anxiolytics and hypnotics, antibiotics, lipid regulators and anti-inflammatories and/or analgesics) were analyzed in both influents and effluents of 15 WWTPs in 5 Portuguese regions (North, Center, Lisbon and Tagus Valley, Alentejo and Algarve), during spring and summer seasons (Pereira et al. 2015). Ciprofloxacin, bezafibrate, gemfibrozil and ibuprofen were quantified at μg L^{-1} levels in both influent and effluent samples, with removal rates ranging from not eliminated to 100 %, depending on the WWTP. The pharmaceuticals quantified at the same levels in influent samples though at ng L^{-1} levels in the effluents were simvastatin, diclofenac and paracetamol. Azithromycin and lorazepam were always detected at ng L^{-1} levels while alprazolam and zolpidem were not found during this study. A monitoring report of the antidepressants citalopram, fluoxetine, paroxetine and sertraline was performed in both influents and effluents of the same WWTPs, during a 1 year follow-up study, embracing the four seasons in 2013 (Silva et al. 2014). In that study, all antidepressants were quantified up to 213.6 ng L^{-1} in the influent samples and up to 95.6 ng L^{-1} in the effluents, with complete removal of fluoxetine and sertraline and a removal of approximately 80 % for citalopram and paroxetine. A group of chiral pharmaceuticals from three different pharmacological classes (antidepressants, beta-blockers and a beta$_2$-adrenergic agonist) including an active metabolite (norfluoxetine) that were quantified in the past as unique molecule entities with no discrimination between enantiomers, were recently studied in wastewaters from three Portuguese WWTP (Ribeiro et al. 2014). Enantiomeric composition was assessed, reinforcing that biodegradation occurring in the secondary treatment of WWTP can be enantioselective and that enantioselectivity depends on the microbial community. In that study, fluoxetine and venlafaxine were detected with higher concentration of (*S*)-fluoxetine and (*S*)-venlafaxine. Concerning beta-blockers, both enantiomers of bisoprolol and propranolol were found in the three WWTPs' effluents and enantiomers of metoprolol were found in two of the WWTPs studied, both under their limits of quantification.

4.2 Personal Care Products

Personal care products such as polycyclic musk fragrances are pollutants widely distributed in the environment. Concern about these compounds is growing due to their potential harmful effects on aquatic organisms and human health (Brausch and Rand 2011). Salgado et al. (2010) reported the presence of polycyclic musks fragrances in 5 WWTPs of Portugal and found their presence in the influent, effluent and secondary sludge between few ng L^{-1} and µg L^{-1}, most frequently in the influent (Table 3). The same group of researchers assessed the variability of polycyclic musks fragrances (Salgado et al. 2011). In that study, the authors demonstrated diurnal variation of mean concentration levels of musks. Lowest values were found at night. These results are important for the planning of sampling campaigns. Machado et al. (2011) detected isoborneol and galaxolide in a wastewater sample and galaxolide in one surface water sample (river Leça) of the eight studied, although groundwater, drinking waters and the other surface waters were free of all compounds included. Galaxolide and tonalide were quantified at similar levels, in two studies regarding one WWTP in the north of Portugal (Salgado et al. 2011; Sousa et al. 2012).

4.3 Portuguese vs *European Wastewaters: Occurrence of PSs and EOPs*

Wastewater samples have been the most studied water matrix concerning the occurrence of pharmaceuticals, with several studies reporting µg L^{-1} levels, revealing the need to improve the actual treatment processes, e.g. through the implementation of advanced oxidation processes (Ribeiro et al. 2015). In fact, various therapeutic classes of pharmaceuticals have been considered however, other classes of pollutants widely investigated in Europe, are still not investigated in wastewaters from Portugal such as EDCs, illicit drugs, pesticides, polycyclic aromatic hydrocarbons, among others (Loos et al. 2012). There is only a brief reference to industrial compounds in an untreated effluent of a textile industry on the outskirts of Porto, being quantified polyethylene glycol 2, polyethoxylated decanol, bis (2-ethylhexyl) phthalate, 3-nitro-BS, 2-NPS and 1-Hydroxy-4-NPS (Farré et al. 2001).

Considering pharmaceuticals, the antiepileptic CBZ was frequently detected, although it was not included in all reports of pharmaceuticals in wastewaters. Antibiotics are a class of contaminants that have been widely investigated, namely fluoroquinolones which were found in the Portuguese wastewaters above 10 µg L^{-1}. These compounds were also found in multi-class studies that have been emerging in the last years, such as the analgesic and antipyretic paracetamol found up to 20 µg L^{-1}, and some NSAIDs, diuretics and lipid regulators found at µg L^{-1} levels (Sousa et al. 2011, 2012; Paíga et al. 2013). The same was verified in other countries. Indeed, the paracetamol, the NSAID ibuprofen and antibiotic drugs are the most abundant in the aquatic environments, with the highest concentrations above several tens of µg L^{-1} in effluents in Greece, Spain and United Kingdom (Jiang et al. 2013). Recently, WWTP effluents from hospitals were also characterized and high levels of iopromide, analgesics and anti-inflammatories, antibiotics, diuretics, antihypertensives and beta-blockers were quantified at μg L^{-1} levels. The same levels were verified in other countries such as Switzerland, France, Spain, among others (Verlicchi et al. 2012; Jiang et al. 2013). Pharmaceutical compounds should be addressed in future studies, correlating the occurrence in different but interconnected matrices, such as wastewaters and surface waters.

The acute negative effects on non-target organism exposed to environmental levels of these substances have been investigated considering parameters such as reproductive and growth rates, stress oxidative response, species mortality beyond others. Despite the high concentrations reported in many studies for WWTP influents and effluents, the values are inferior to the concentration which gives 50 % of the maximum response (EC_{50}) considering species commonly used in environmental toxicology studies, such as fish, *Daphnia* and algae (Ginebreda et al. 2010). Nevertheless, the chronic expose to biological active substances even at low levels represents a threat to wildlife and human health. Bioaccumulation and chronic tests are very scarce and thus, the extent of the consequences is still poorly understood. Furthermore, it is important to evaluate the toxicological effects of mixtures of these substances. Most compounds are found as part of mixtures, which individually may interact with each other to trigger negative effects. As a consequence, the concentration of a target compound giving no observed adverse effects may underestimate the risk if analysed as a single substance. Thus, the development of toxicological studies is important to the establishment of a list of different classes of pharmaceutical compounds that should be considered in systematic studies, as well as to define their maximum concentration in future environmental regulations.

There is only one Portuguese study on quantification of chiral pharmaceuticals from different pharmacological classes, including a metabolite, in wastewaters from three WWTPs (Ribeiro et al. 2014). Despite the well recognize importance of chirality in pharmaceuticals, pesticides, fragrances and many other fields, most of the environmental studies neglect the existence of enantiomers. Biodegradation and adsorption to biomass occurring in the secondary treatment of WWTPs can be enantioselective and the enrichment of one enantiomer or other phenomena as racemisation or enantiomerization occur frequently (Ribeiro et al. 2012). Considering the differences in the toxicity of the enantiomers, it is crucial to assess the variation of the enantiomeric fraction during the processes occurring at WWTPs as well as in other aquatic compartments, to further understand the effects of enantiomers in the ecosystems (Ribeiro et al. 2012).

Few studies comprising personal care products in the Portuguese wastewaters have found polycyclic musks fragrances between ng L^{-1} and μg L^{-1}. These compounds were only studied in wastewater matrices and should be further studied in other aquatic matrices. These pollutants have been found in the same concentration order in France, Switzerland, Spain, among others (Verlicchi et al. 2012; Jiang et al. 2013).

Unlike surface waters, there are few studies on industrial compounds in wastewaters, being imperative the characterization of these kind of compounds in wastewaters since they are pointed out in the Directive 2013/39/EU

(EU-Directive 2013). In fact, PSs such as BPA, APs and APEOs have not been investigated in wastewaters in Portugal, however they have been found in various countries in the order of μg L^{-1} (Bergé et al. 2012).

Considering the data here reviewed and the concentrations reported in the WWTPs for a great number of substances not investigated yet in receiving surface waters, simulation models to evaluate the influence of WWTP discharges on river waters are needed in Portugal. For example, the GREAT-ER model (Geography-referenced Regional Exposure Assessment Tool for European Rivers) was already used for modelling pollutants in European catchments (Alder et al. 2010; Aldekoa et al. 2013), giving information about aquatic exposure assessment and source variability effect on river water quality. However, there are no reports in Portugal that implement this model to river basins influenced by discharges of WWTPs.

5 Summary

Studies on occurrence of PSs and EOPs in Portugal are in agreement with the reports of other countries, reinforcing the need for risk assessment through monitoring programs and toxicity evaluation for aquatic animals and humans. Even compounds already banned in Europe, included in the PSs list by the WFD such as atrazine, APs, APEs, are still found in various aquatic samples. Estrogens and pharmaceuticals are compounds of recent concern, regarding the regulatory approach, which includes only EE2, diclofenac and macrolide antibiotics. These compounds were recommended to be part of the first PSs watch list. A wide range of pesticides classes has been studied in Portuguese surface waters and groundwaters. Almost all classes of pesticides (e.g., OCPs, herbicides, OPPs, pyrethroids and triazines) were found, indicating that these compounds are widely used in the Portuguese territory. Most of them followed a seasonal pattern and particular attention should be given to regions close to agricultural areas. Nonetheless, occurrence of pesticides in surface waters should be more investigated. Regarding WWTPs, most occurrence studies are focused on pharmaceuticals, which have been found at worrying μg L^{-1} levels; however other classes of substances such as EDCs, illicit drugs, pesticides, polycyclic aromatic hydrocarbons, many of them defined as PSs were not determined in wastewaters yet. Since WWTPs discharge their effluents into rivers, where water is then extracted for the production of drinking water, it is very important to continue the research in this area, widening to more classes of compounds and including the receiving surface waters. It is also imperative to develop monitoring and systematic studies that will allow not only assessing the presence and concentration of the pollutants, but also evaluate spatiotemporal distribution, their ecotoxicological relevance and their possible sources in the aquatic systems. Simulation models should be included in future monitoring programs to compare with sampling data to better adjust the models, with the main goal of assessing emissions and degradation of the pollutants. Thereby, providing valuable information to assess the pathway of these pollutants in the environment, may allow proposing the adaptation of more suitable removal methods at the existing WWTPs or the improvements in their future design.

Acknowledgements The authors wish to acknowledge the support from CESPU (03-GCQF-CICS-11) PHARMADRUGS-CESPU-2014 and Fundação para a Ciência e Tecnologia (FCT) through the projects FLUOROPHARMA (PTDC/EBB-EBI/111699/2009) and Strategic Funding UID/Multi/04423/2013 through national funds provided by FCT and European Regional Development Fund (ERDF), in the framework of the programme PT2020. Ana Rita Ribeiro wishes to acknowledge the research grant from FCT, Portugal (Ref. SFRH/BPD/101703/2014) and the support from the project NORTE-07-0202-FEDER-038900 (NEPCAT), financed by FEDER through ON2 (Programa Operacional do Norte) and QREN.

References

Aldekoa J, Medici C, Osorio V, Pérez S, Marcé R, Barceló D, Francés F (2013) Modelling the emerging pollutant diclofenac with the GREAT-ER model: application to the Llobregat River Basin. J Hazard Mater 263(Part 1):207–213

Alder AC, Schaffner C, Majewsky M, Klasmeier J, Fenner K (2010) Fate of β-blocker human pharmaceuticals in surface water: comparison of measured and simulated concentrations in the Glatt Valley Watershed, Switzerland. Water Res 44:936–948

Andrade AIASS, Stigter TY (2009) Multi-method assessment of nitrate and pesticide contamination in shallow alluvial groundwater as a function of hydrogeological setting and land use. Agr Water Manag 96(12):1751–1765

Azevedo DD, Lacorte S, Barcelo PV, Barcelo D (2001) Occurrence of nonylphenol and bisphenol-A in surface waters from Portugal. J Braz Chem Soc 12(4):532–537

Azevedo DD, Gerchon E, dos Reis EO (2004) Monitoring of pesticides and polycyclic aromatic hydrocarbons in water from Paraiba do Sul River, Brazil. J Braz Chem Soc 15(2):292–299

Azevedo IC, Duarte PM, Bordalo AA (2006) Pelagic metabolism of the Douro estuary (Portugal) - factors controlling primary production. Estuar Coast Shelf Sci 69(1-2):133–146

Bacaloni A, Cavaliere C, Faberi A, Foglia P, Samperi R, Lagana A (2005) Determination of isoflavones and coumestrol in river water and domestic wastewater sewage treatment plants. Anal Chim Acta 531(2):229–237

Baran N, Mouvet C, Négrel P (2007) Hydrodynamic and geochemical constraints on pesticide concentrations in the groundwater of an agricultural catchment (Brévilles, France). Environ Pollut 148(3):729–738

Baronti C, Curini R, D'Ascenzo G, Di Corcia A, Gentili A, Samperi R (2000) Monitoring natural and synthetic estrogens at activated sludge sewage treatment plants and in a receiving river water. Environ Sci Technol 34(24):5059–5066

Belfroid AC, Van der Horst A, Vethaak AD, Schafer AJ, Rijs GBJ, Wegener J, Cofino WP (1999) Analysis and occurrence of estrogenic hormones and their glucuronides in surface water and waste water in The Netherlands. Sci Total Environ 225(1-2):101–108

Bergé A, Cladière M, Gasperi J, Coursimault A, Tassin B, Moilleron R (2012) Meta-analysis of environmental contamination by alkylphenols. Environ Sci Pollut Res 19:3798–3819

Blair BD, Crago JP, Hedman CJ, Klaper RD (2013) Pharmaceuticals and personal care products found in the Great Lakes above concentrations of environmental concern. Chemosphere 93 (9):2116–2123

Bordalo AA, Vieira MEC (2005) Spatial variability of phytoplankton, bacteria and viruses in the mesotidal salt wedge Douro Estuary (Portugal). Estuar Coast Shelf Sci 63(1-2):143–154

Bordalo AA, Teixeira R, Wiebe WJ (2006) A water quality index applied to an international shared river basin: the case of the Douro river. Environ Manage 38:910–920

Boyd GR, Reemtsma H, Grimm DA, Mitra S (2003) Pharmaceuticals and personal care products (PPCPs) in surface and treated waters of Louisiana, USA and Ontario, Canada. Sci Total Environ 311(1–3):135–149

Braganca I, Placido A, Paiga P, Domingues VF, Delerue-Matos C (2012) QuEChERS: a new sample preparation approach for the determination of ibuprofen and its metabolites in soils. Sci Total Environ 433:281–289

Brausch JM, Rand GM (2011) A review of personal care products in the aquatic environment: environmental concentrations and toxicity. Chemosphere 82(11):1518–1532

Caeiro S, Costa MH, Goovaerts P, Martins F (2005) Benthic biotope index for classifying habitats in the sado estuary: Portugal. Mar Environ Res 60:570–593

Calisto V, Bahlmann A, Schneider RJ, Esteves VI (2011) Application of an ELISA to the quantification of carbamazepine in ground, surface and wastewaters and validation with LC-MS/MS. Chemosphere 84(11):1708–1715

Cargouet M, Perdiz D, Mouatassim-Souali A, Tamisier-Karolak S, Levi Y (2004) Assessment of river contamination by estrogenic compounds in Paris area (France). Sci Total Environ 324 (1-3):55–66

Cargouet M, Perdiz D, Levi Y (2007) Evaluation of the estrogenic potential of river and treated waters in the Paris area (France) using in vivo and in vitro assays. Ecotoxicol Environ Saf 67 (1):149–156

Carmona E, Andreu V, Picó Y (2014) Occurrence of acidic pharmaceuticals and personal care products in Turia River Basin: from waste to drinking water. Sci Total Environ 484:53–63

Carvalho JJ, Jerónimo PCA, Gonçalves C, Alpendurada MF (2008) Evaluation of a multiresidue method for measuring fourteen chemical groups of pesticides in water by use of LC-MS-MS. Anal Bioanal Chem 392(5):955–968

Cerejeira MJ, Viana P, Batista S, Pereira T, Silva E, Valério MJ, Silva A, Ferreira M, Silva-Fernandes AM (2003) Pesticides in Portuguese surface and ground waters. Water Res 37 (5):1055–1063

da Silva BF, Jelic A, López-Serna R, Mozeto AA, Petrovic M, Barceló D (2011) Occurrence and distribution of pharmaceuticals in surface water, suspended solids and sediments of the Ebro river basin, Spain. Chemosphere 85(8):1331–1339

Díez S, Lacorte S, Viana P, Barceló D, Bayona JM (2005) Survey of organotin compounds in rivers and coastal environments in Portugal 1999–2000. Environ Pollut 136(3):525–536

Echeandía JS, Vale C, Caetano M, Pereira P, Prego R (2010) Effect of tidal flooding on metal distribution in pore waters of marsh sediments and its transport to water column (Tagus estuary, Portugal). Mar Environ Res 70:358–367

EPAL (2012) Relatório 2011-2012 - Qualidade da Água para Consumo Humano

Erbs M, Hoerger CC, Hartmann N, Bucheli TD (2007) Quantification of six phytoestrogens at the nanogram per liter level in aqueous environmental samples using C-13(3)-labeled internal standards. J Agric Food Chem 55(21):8339–8345

Esteban S, Gorga M, Petrovic M, González-Alonso S, Barceló D, Valcárcel Y (2014) Analysis and occurrence of endocrine-disrupting compounds and estrogenic activity in the surface waters of Central Spain. Sci Total Environ 466–467:939–951

EU-Decision (2001) Commission Decision (2455/2001/EC) of the European Parliament and of the Council establishing the list of priority substances in the field of water policy and amending Directive 2000/60/EC. Official Journal of the European Communities L 331/1

EU-Decision (2015) (2015/495) of 20 March 2015 establishing a watch list of substances for Union-wide monitoring in the field of water policy pursuant to Directive 2008/105/EC of the European Parliament and of the Council. Official Journal of the European Communities L 78/40

EU-Directive (2000) Directive of the European Parliament and of the Council 2000/60/EC, establishing a framework for community action in the field of water policy. Official Journal of the European Communities L 327:1–72

EU-Directive (2008) Directive 2008/105/EC of the European Parliament and of Council on environmental quality standard in the field of water policy, amending and subsequently repealing Council Directives 82/176/EEC, 83/513/EEC, 84/156/EEC, 84/491/EEC and amending Directive 2000/60/EC of the European Parliament and of the Council. Official Journal of the European Communities L 348:92–93

EU-Directive (2013) DIRECTIVE 2013/39/EU OF THE EUROPEAN PARLIAMENT AND OF THE COUNCIL of 12 August 2013 amending Directives 2000/60/EC and 2008/105/EC as regards priority substances in the field of water policy. Official Journal of the European Union L 226:1–17

Farré M, Pasini O, Carmen Alonso M, Castillo M, Barceló D (2001) Toxicity assessment of organic pollution in wastewaters using a bacterial biosensor. Anal Chim Acta 426(2):155–165

Farré M, Gonçalves C, Lacorte S, Barceló D, Alpendurada M (2002) Pesticide toxicity assessment using an electrochemical biosensor with Pseudomonas putida and a bioluminescence inhibition assay with Vibrio fischeri. Anal Bioanal Chem 373(8):696–703

Gaffney VJ, Almeida CMM, Rodrigues A, Ferreira E, Benoliel MJ, Cardoso VC (2015) Occurrence of pharmaceuticals in a water supply system and related human health risk assessment. Water Res 72:199–208

Gil O, Vale C (2001) Evidence for polychlorinated biphenyls dechlorination in the sediments of Sado Estuary, Portugal. Mar Pollut Bull 42(6):452–460

Ginebreda A, Muñoz I, de Alda ML, Brix R, López-Doval J, Barceló D (2010) Environmental risk assessment of pharmaceuticals in rivers: relationships between hazard indexes and aquatic macroinvertebrate diversity indexes in the Llobregat River (NE Spain). Environ Int 36 (2):153–162

Gonçalves CM, Esteves da Silva JCG, Alpendurada MF (2007) Evaluation of the pesticide contamination of groundwater sampled over two years from a vulnerable zone in Portugal. J Agric Food Chem 55(15):6227–6235

Gros M, Rodríguez-Mozaz S, Barceló D (2012) Fast and comprehensive multi-residue analysis of a broad range of human and veterinary pharmaceuticals and some of their metabolites in surface and treated waters by ultra-high-performance liquid chromatography coupled to quadrupole-linear ion trap tandem mass spectrometry. J Chromatogr A 1248:104–121

Jiang J-Q, Zhou Z, Sharma VK (2013) Occurrence, transportation, monitoring and treatment of emerging micro-pollutants in waste water—a review from global views. Microchem J 110:292–300

Jobling S, Tyler CR, Sumpter JP, Nolan M, Brighty G (1998) Widespread sexual disruption in wild fish. Environ Sci Tech 32(17):2498–2506

Jobling S, Williams R, Johnson A, Taylor A, Gross-Sorokin M, Nolan M, Tyler CR, van Aerle R, Santos E, Brighty G (2006) Predicted exposures to steroid estrogens in UK rivers correlate with widespread sexual disruption in wild fish populations. Environ Health Perspect 114:32–39

Jurado A, Vàzquez-Suñé E, Carrera J, López de Alda M, Pujades E, Barceló D (2012) Emerging organic contaminants in groundwater in Spain: a review of sources, recent occurrence and fate in a European context. Sci Total Environ 440:82–94

Kim SD, Cho J, Kim IS, Vanderford BJ, Snyder SA (2007) Occurrence and removal of pharmaceuticals and endocrine disruptors in South Korean surface, drinking, and waste waters. Water Res 41(5):1013–1021

Lagana A, Bacaloni A, De Leva I, Faberi A, Fago G, Marino A (2004) Analytical methodologies for determining the occurrence of endocrine disrupting chemicals in sewage treatment plants and natural waters. Anal Chim Acta 501(1):79–88

Lin AY-C, Tsai Y-T (2009) Occurrence of pharmaceuticals in Taiwan's surface waters: impact of waste streams from hospitals and pharmaceutical production facilities. Sci Total Environ 407 (12):3793–3802

Liu J-L, Wong M-H (2013) Pharmaceuticals and personal care products (PPCPs): a review on environmental contamination in China. Environ Int 59:208–224

Loos R, Carvalho R, Comero S, António DC, Ghiani M, Lettieri T, Locoro G, Paracchini B, Tavazzi S, Gawlik BM, Blaha L, Jarosova B, Voorspoels S, Schwesig D, Haglund P, Fick J, Gans O (2012). EU Wide Monitoring Survey on Waste Water Treatment Plant Effluents. JRC Scientific and Policy Reports. Available at http://www.jrc.ec.europa.eu/

Machado S, Gonçalves C, Cunha E, Guimarães A, Alpendurada MF (2011) New developments in the analysis of fragrances and earthy–musty compounds in water by solid-phase microextraction (metal alloy fibre) coupled with gas chromatography–(tandem) mass spectrometry. Talanta 84(4):1133–1140

Madureira TV, Barreiro JC, Rocha MJ, Rocha E, Cass QB, Tiritan ME (2010) Spatiotemporal distribution of pharmaceuticals in the Douro River estuary (Portugal). Sci Total Environ 408 (22):5513–5520

Meffe R, Bustamante I (2014) Emerging organic contaminants in surface water and groundwater: a first overview of the situation in Italy. Sci Total Environ 481:280–295

Morvan X, Mouvet C, Baran N, Gutierrez A (2006) Pesticides in the groundwater of a spring draining a sandy aquifer: temporal variability of concentrations and fluxes. J Contam Hydrol 87 (3–4):176–190

Mucha AP, Bordalo AA, Vasconcelos M (2004) Sediment quality in the Douro river estuary based on trace metal contents, macrobenthic community and elutriate sediment toxicity test (ESTT). J Environ Monit 6(7):585–592

Noppe H, Verslycke T, De Wulf E, Verheyden K, Monteyne E, Van Caeter P, Janssen CR, De Brabander HF (2007) Occurrence of estrogens in the Scheldt estuary: a 2-year survey. Ecotoxicol Environ Saf 66(1):1–8

Nunes BS, Rodrigues S, Antunes SC, Castro BB, Gonçalves F (2011a) The use of biomarkers to assess aquatic pollution by widespread agents (detergents – SDS; pesticides – chlorfenvinphos): feasibility, responsiveness and biological consequences in fish. Toxicol Lett 205:S67

Nunes M, Marchand P, Vernisseau A, Bizec BL, Ramos F, Pardal MA (2011b) PCDD/Fs and dioxin-like PCBs in sediment and biota from the Mondego estuary (Portugal). Chemosphere 83 (10):1345–1352

Osorio V, Marcé R, Pérez S, Ginebreda A, Cortina JL, Barceló D (2012) Occurrence and modeling of pharmaceuticals on a sewage-impacted Mediterranean river and their dynamics under different hydrological conditions. Sci Total Environ 440:3–13

Paíga P, Santos LMLM, Amorim C, Araújo A, Montenegro MCSM, Pena A, Delerue-Matos C (2013) Pilot monitoring study of ibuprofen in surface waters of north of Portugal. Environ Sci Pollut Res 20(4):2410–2420

Pena A, Chmielova D, Lino CM, Solich P (2007) Determination of Fluoroquinolone antibiotics in surface waters from Mondego River by high performance liquid chromatography using monolithic column. J Sep Sci 30:2924–2928

Pereira P, Pablo H, Subida MD, Vale C, Pacheco M (2009) Biochemical responses of the shorecrab (Carcinus maenas) in a eutrophic and metal-contaminated coastal system (Óbidos lagoon, Portugal). Ecotoxicol Environ Saf 72:1471–1480

Pereira AMPT, Silva LJG, Meisel LM, Lino CM, Pena A (2015) Environmental impact of pharmaceuticals from Portuguese wastewaters: geographical and seasonal occurrence, removal and risk assessment. Environ Res 136:108–119

Piersma AH, Luijten M, Popov V, Tomenk V, Altstein M, Kagampang F, Schlesinger H (2009) Endocrine-disrupting chemicals in food. In: Shaw I (ed) Pharmaceuticals. University of Canterbury, Christchurch, pp 459–518

Postigo C, Barceló D (2015) Synthetic organic compounds and their transformation products in groundwater: occurrence, fate and mitigation. Sci Total Environ 503–504:32–47. doi:10.1016/j.scitotenv.2014.06.019

Quevauviller P, Lavigne R, Pinel R, Astruc M (1989) Organo-tins in sediments and mussels from the Sado estuarine system (Portugal). Environ Pollut 57(2):149–166

Quiros L, Cespedes R, Lacorte S, Viana P, Raldua D, Barcelo D, Pina B (2005) Detection and evaluation of endocrine-disruption activity in water samples from Portuguese rivers. Environ Toxicol Chem 24(2):389–395

Ramalhosa E, Pereira E, Vale C, Valega M, Monterroso P, Duarte AC (2005) Mercury distribution in Douro estuary (Portugal). Mar Pollut Bull 50(11):1218–1222

Rato M, Sousa A, Quinta R, Langston W, Barroso C (2006) Assessment of inshore/offshore tributyltin pollution gradients in the northwest Portugal continental shelf using Nassarius reticulatus as a bioindicator. Environ Toxicol Chem 25(12):3213–3220

Ribeiro C, Tiritan ME, Rocha E, Rocha MJ (2009a) Seasonal and spatial distribution of several endocrine-disrupting compounds in the Douro River Estuary, Portugal. Arch Environ Contam Toxicol 56(1):1–11

Ribeiro C, Pardal MA, Tiritan ME, Rocha E, Margalho R, Rocha MJ (2009b) Spatial distribution and quantification of endocrine disrupting chemicals in the Sado River estuary, Portugal. Environ Monit Assess 159:415–427

Ribeiro C, Pardal MA, Martinho F, Margalho R, Tiritan ME, Rocha E, Rocha MJ (2009c) Distribution of endocrine disruptors in the Mondego River estuary, Portugal. Environ Monit Assess 149:183–193

Ribeiro AR, Castro PML, Tiritan ME (2012) Chiral pharmaceuticals in the environment. Environ Chem Lett 10:239–253

Ribeiro AR, Santos LHMLM, Maia AS, Delerue-Matos C, Castro PML, Tiritan ME (2014) Enantiomeric fraction evaluation of pharmaceuticals in environmental matrices by liquid chromatography-tandem mass spectrometry. J Chromatogr A 1363:226. doi:10.1016/j.chroma.2014.06.099

Ribeiro AR, Nunes OC, Pereira MFR, Silva AMT (2015) An overview on the advanced oxidation processes applied for the treatment of water pollutants defined in the recently launched Directive 2013/39/EU. Environ Int 75:33–51

Rocha MJ, Ribeiro C, Ribeiro M (2011) Development and optimisation of a GC-MS method for the evaluation of oestrogens and persistent pollutants in river and seawater samples. Int J Environ Anal Chem 91(12):1191–1205

Rocha MJ, Cruzeiro C, Ferreira C, Rocha E (2012a) Occurrence of endocrine disruptor compounds in the estuary of the Iberian Douro River and nearby Porto Coast (NW Portugal). Toxicol Environ Chem 94(2):252–261

Rocha MJ, Cruzeiro C, Reis M, Rocha E, Pardal MA (2012b) Determination of 17 endocrine disruptor compounds and their spatial and seasonal distribution in the Sado River Estuary (Portugal). Toxicol Environ Chem 95(2):237–253

Rocha MJ, Ribeiro M, Ribeiro C, Couto C, Cruzeiro C, Rocha E (2012c) Endocrine disruptors in the Leça River and nearby Porto Coast (NW Portugal): presence of estrogenic compounds and hypoxic conditions. Toxicol Environ Chem 94(2):262–274

Rocha MJ, Ribeiro M, Cruzeiro C, Figueiredo F, Rocha E (2012d) Development and validation of a GCMS method for determination of 39 common pesticides in estuarine water – targeting hazardous amounts in the Douro River estuary. Int J Environ Anal Chem 92:1587–1608

Rocha MJ, Cruzeiro C, Rocha E (2013a) Quantification of 17 endocrine disruptor compounds and their spatial and seasonal distribution in the Iberian Ave River and its coastline. Toxicol Environ Chem 95(3):386–399

Rocha MJ, Cruzeiro C, Rocha E (2013b) Development and validation of a GC-MS method for the evaluation of 17 endocrine disruptor compounds, including phytoestrogens and sitosterol, in coastal waters - their spatial and seasonal levels in Porto costal region (Portugal). J Water Health 11(2):281–296

Rocha MJ, Cruzeiro C, Reis M, Rocha E, Pardal M (2013c) Determination of seventeen endocrine disruptor compounds and their spatial and seasonal distribution in Ria Formosa Lagoon (Portugal). Environ Monit Assess 185(10):8215–8226

Rodrigues ET, Lopes I, Pardal MÂ (2013) Occurrence, fate and effects of azoxystrobin in aquatic ecosystems: a review. Environ Int 53:18–28

Salgado R, Noronha JP, Oehmen A, Carvalho G, Reis MA (2010) Analysis of 65 pharmaceuticals and personal care products in 5 wastewater treatment plants in Portugal using a simplified analytical methodology. Water Sci Tech 62(12):2862–2871

Salgado R, Marques R, Noronha JP, Mexia JT, Carvalho G, Oehmen A, Reis MAM (2011) Assessing the diurnal variability of pharmaceutical and personal care products in a full-scale activated sludge plant. Environ Pollut 159(10):2359–2367

Salgado R, Marques R, Noronha JP, Carvalho G, Oehmen A, Reis MAM (2012) Assessing the removal of pharmaceuticals and personal care products in a full-scale activated sludge plant. Environ Sci Pollut Res 19(5):1818–1827

Santos LHMLM, Gros M, Rodriguez-Mozaz S, Delerue-Matos C, Pena A, Barceló D, Montenegro MCBSM (2013a) Contribution of hospital effluents to the load of pharmaceuticals in urban wastewaters: identification of ecologically relevant pharmaceuticals. Sci Total Environ 461–462:302–316

Santos LHMLM, Paíga P, Araújo AN, Pena A, Delerue-Matos C, Montenegro MCBSM (2013b) Development of a simple analytical method for the simultaneous determination of paracetamol, paracetamol-glucuronide and p-aminophenol in river water. J Chromatogr B 930:75–81

Schriks M, Heringa MB, van der Kooi MME, de Voogt P, van Wezel AP (2010) Toxicological relevance of emerging contaminants for drinking water quality. Water Res 44(2):461–476

Schwartz P, Thorpe KL, Bucheli TD, Wettstein FE, Burkhardt-Holm P (2010) Short-term exposure to the environmentally relevant estrogenic mycotoxin zearalenone impairs reproduction in fish. Sci Total Environ 409(2):326–333

Seifrtová M, Pena A, Lino CM, Solich P (2008) Determination of fluoroquinolone antibiotics in hospital and municipal wastewaters in Coimbra by liquid chromatography with a monolithic column and fluorescence detection. Anal Bioanal Chem 391(3):799–805

Silva E, Batista S, Viana P, Antunes P, Serodio L, Cardoso AT, Cerejeira MJ (2006) Pesticides and nitrates in groundwater from oriziculture areas of the 'Baixo Sado' region (Portugal). Int J Environ Anal Chem 86(13):955–972

Silva LJG, Pereira AMPT, Meisel LM, Lino CM, Pena A (2014) A one-year follow-up analysis of antidepressants in Portuguese wastewaters: occurrence and fate, seasonal influence, and risk assessment. Sci Total Environ 490:279–287

Soares H, Boaventura RAR, Machado A, da Silva J (1999) Sediments as monitors of heavy metal contamination in the Ave river basin (Portugal): multivariate analysis of data. Environ Pollut 105(3):311–323

Sousa MA, Gonçalves C, Cunha E, Hajšlová J, Alpendurada MF (2011) Cleanup strategies and advantages in the determination of several therapeutic classes of pharmaceuticals in wastewater samples by SPE–LC–MS/MS. Anal Bioanal Chem 399(2):807–822

Sousa MA, Gonçalves C, Vilar VJP, Boaventura RAR, Alpendurada MF (2012) Suspended TiO2-assisted photocatalytic degradation of emerging contaminants in a municipal WWTP effluent using a solar pilot plant with CPCs. Chem Eng J 198–199:301–309

Vasconcelos RP, Reis-Santos P, Fonseca V, Maia A, Ruano M, Franca S, Vinagre C, Costa MJ, Cabral H (2007) Assessing anthropogenic pressures on estuarine fish nurseries along the Portuguese coast: a multi-metric index and conceptual approach. Sci Total Environ 374 (2-3):199–215

Verlicchi P, Aukidy MA, Zambello E (2012) Occurrence of pharmaceutical compounds in urban wastewater: removal, mass load and environmental risk after a secondary treatment—a review. Sci Total Environ 429:123–155

Villaverde J, Hildebrandt A, Martínez E, Lacorte S, Morillo E, Maqueda C, Viana P, Barceló D (2008) Priority pesticides and their degradation products in river sediments from Portugal. Sci Total Environ 390:507–513

Vitali M, Ensabella F, Stella D, Guidotti M (2004) Nonylphenols in freshwaters of the hydrologic system of an Italian district: association with human activities and evaluation of human exposure. Chemosphere 57(11):1637–1647

Wang DF, Zhang NY, Peng YZ, Qi DS (2010) Interaction of zearalenone and soybean isoflavone on the development of reproductive organs, reproductive hormones and estrogen receptor expression in prepubertal gilts. Anim Reprod Sci 122(3-4):317–323

Waring RH, Harris RM (2005) Endocrine disrupters: a human risk? Mol Cell Endocrinol 244 (1-2):2–9

Wennmalm Å (2011) Pharmaceuticals: environmental effects. In: Jerome ON (ed) Encyclopedia of environmental health. Elsevier, Burlington, MA, pp 462–471

Werner J, Wautier K, Evans RE, Baron CL, Kidd K, Palace V (2003) Waterborne ethynylestradiol induces vitellogenin and alters metallothionein expression in lake trout (Salvelinus namaycush). Aquat Toxicol 62(4):321–328

Wille K, De Brabander HF, Vanhaecke L, De Wulf E, Van Caeter P, Janssen CR (2012) Coupled chromatographic and mass-spectrometric techniques for the analysis of emerging pollutants in the aquatic environment. Trends Anal Chem 35:87–108

Writer JH, Ferrer I, Barber LB, Thurman EM (2013) Widespread occurrence of neuro-active pharmaceuticals and metabolites in 24 Minnesota rivers and wastewaters. Sci Total Environ 461–462:519–527

Wu J-P, Luo X-J, Zhang Y, Luo Y, Chen S-J, Mai B-X, Yang Z-Y (2008) Bioaccumulation of polybrominated diphenyl ethers (PBDEs) and polychlorinated biphenyls (PCBs) in wild aquatic species from an electronic waste (e-waste) recycling site in South China. Environ Int 34(8):1109–1113

Xue X, Wu F, Zhang X, Deng N (2008) Occurrence of endocrine disrupting compounds in rivers and lakes of Wuhan City, China. Fresenius Environ Bull 17(2):203–210

Ying G-G, Williams B, Kookana R (2002) Environmental fate of alkylphenols and alkylphenol ethoxylates--a review. Environ Int 28(3):215–226

Yokota H, Tsuruda Y, Maeda M, Oshima Y, Tadokoro H, Nakazono A, Honjo T, Kobayashi K (2000) Effect of bisphenol A on the early life stage in Japanese medaka (Oryzias latipes). Environ Toxicol Chem 19(7):1925–1930

Yoon Y, Ryu J, Oh J, Choi B-G, Snyder SA (2010) Occurrence of endocrine disrupting compounds, pharmaceuticals, and personal care products in the Han River (Seoul, South Korea). Sci Total Environ 408(3):636–643

Zhan M, Yang X, Xian Q, Kong L (2006) Photosensitized degradation of bisphenol A involving reactive oxygen species in the presence of humic substances. Chemosphere 63(3):378–386

Zhang Y, Geißen S-U, Gal C (2008) Carbamazepine and diclofenac: removal in wastewater treatment plants and occurrence in water bodies. Chemosphere 73(8):1151–1161

Zhou Q, Zhang J, Fu J, Shi J, Jiang G (2008a) Biomonitoring: an appealing tool for assessment of metal pollution in the aquatic ecosystem. Anal Chim Acta 606(2):135–150

Zhou R, Zhu L, Chen Y, Kong Q (2008b) Concentrations and characteristics of organochlorine pesticides in aquatic biota from Qiantang River in China. Environ Pollut 151(1):190–199

Zuo YG, Zhang K, Deng YW (2006) Occurrence and photochemical degradation of 17 alpha-ethinylestradiol in Acushnet River Estuary. Chemosphere 63(9):1583–1590

Lead Toxicity, Antioxidant Defense and Environment

Alessia Carocci, Alessia Catalano, Graziantonio Lauria, Maria Stefania Sinicropi, and Giuseppe Genchi

Contents

A. Carocci • A. Catalano
Dipartimento di Farmacia-Scienze del Farmaco, Università degli Studi di Bari "A. Moro", 70125 Bari, Italy

G. Lauria • M.S. Sinicropi (✉) • G. Genchi (✉)
Dipartimento di Farmacia e Scienze della Salute e della Nutrizione, Università della Calabria, 87036 Arcavacata di Rende, Cosenza, Italy
e-mail: s.sinicropi@unical.it; giuseppe.genchi@unical.it

P. de Voogt (ed.), *Reviews of Environmental Contamination and Toxicology, Volume 238*, Reviews of Environmental Contamination and Toxicology 238,
DOI 10.1007/398_2015_5003

1 Introduction

Environmental and occupational exposure to a large number of chemicals occurs at various stages throughout human life. Many of these are devoid of toxicity, but some could pose a significant health risk, i.e. the exposure to environmental xenobiotic metals as lead, mercury (Sinicropi et al. 2010a; Carocci et al. 2014), cadmium, etc. In particular, lead has long been a widespread public concern (Basha and Reddy 2010). Lead is one of the earliest heavy metals discovered by men. Due to its unique properties, as low melting point, softness, malleability, ductility, and resistance to corrosion, men have used lead for the last 5000 years in a wide range of applications.

Ancient civilizations have used the lead for the manufacture of kitchen utensils and decorative articles, in plumbing, tableware, and other products since the Roman Empire. It has been utilized in pipes, pigments and paints, construction materials, glass, ceramics, too. Successively, lead has been used as anti-knock fuel additive, in lead-acid batteries, electronic components, and to a lesser extent in ammunition, cable coverings, some paints and ceramics, and in soldering in the food canning industries. It has even been used in some medicines and cosmetics.

The non-biodegradable nature of lead is the reason for its prolonged persistence in the soil, air, and drinking water. The use of lead in pipes, paints, and gasoline additives resulted in large amounts of lead entering the environment. It is a multimedia pollutant, since the human exposure occurs via inhaled air, dust, food and drinking water. However, lead has no known biological function in humans.

In the last 30 years, this metal has been reduced or eliminated in most gasoline and paints with a concurrent drop of blood lead levels of adults and children. However, there are still lead sources near incinerators and smelters, and in poor city areas where peeling Pb-based paint is a source of exposure to children.

Lead accumulates in bone, with clearance half times of approximately two to three decades from cortical bone (e.g., tibia), where it can be measured using X-ray fluorescence. It can contribute to blood lead levels, although the latter are primarily influenced by new external exposure. Blood lead level is the best available estimate of recent dose, while tibia lead is an estimate of lifetime retained cumulative dose (Somervaille et al. 1988; Gulson 2000).

After its accumulation in the bone, lead is released over time especially during periods of bone demineralization such as pregnancy, lactation and postmenopause. Therefore, mobilization of lead from bone can occur in physiological or pathological situations, entailing an increased bone turnover or demineralization as, for instance, in osteoporosis, bone fracture, pregnancy, hyperthyroidism, bone cancer, and chemotherapy (Silbergeld et al. 1988; Silbergeld 1991).

Acute toxicity is related to occupational intense exposure of short duration reaching blood lead levels of 100–120 μg/dl. On the other hand, chronic toxicity is much more common and occurs after repeated exposure over a prolonged period, blood lead levels being of 40–60 μg/dl. If not treated in time, persistent vomiting, encephalopathy, lethargy, convulsions, delirium, and coma can occur (Flora et al. 2006; Pearce 2007).

Numerous studies regarding the toxicology of lead have shown it to be a potent neurotoxicant, especially during nervous system development (Feldman et al. 1980). High levels lead exposure in adults and children has been associated with deficits in memory and in intellectual functioning (Arvig et al. 1980; Baghurst et al. 1992; Baker 1982; Baker et al. 1984), attention and concentration (Arvig et al. 1980; Stollery et al. 1989), speed and psychomotor performance (Arvig et al. 1980; Stollery et al. 1991). Lead has also well-known effects on the cardio-vascular (Vaziri 2002), renal (Gonick 2002), reproductive (Bellinger 2005) and immune (Dietert and Piepenbrink 2006) systems, as well as bones and teeth (Hu et al. 1998) and it has been also identified as a probable human carcinogen (Silbergeld 2003; van Wijngaarden and Dosemeci 2006). Epidemiologic studies suggest an association of inorganic lead exposure to lung, stomach and, to a lesser extent, kidney and brain cancer (Steenland and Boffetta 2000; International Agency for Research in Cancer, IARC 2006).

Lead is known to disrupt dopaminergic function in experimental studies; it seems also to induce oxidative stress (Ercal et al. 2001), which is a candidate hypothesis for the etiology of Parkinson, Alzheimer and other neurodegenerative age-related diseases.

2 Chemical Form and Properties of Lead

Lead (Pb) is a bluish gray heavy metal (atomic weight 207.2), that occurs naturally in various mineral forms in the earth's crust. Metallic lead is resistant to corrosion, because, when it is exposed to air or water, thin films of lead compounds (oxides and carbonates) are formed and protect this metal from further attacks. It has been widely used for centuries because it is readily shaped, molded, and resistant to corrosion. Lead can exist in three forms: metallic, inorganic, and organic. Lead in the environment rarely occurs in its elemental state, but rather in its oxidation state (Pb^{2+}) in various ores throughout the earth. The phasing out of leaded gasoline for transportation vehicles between 1973 and 1995 and the removal of lead from paint by 1978 have resulted in substantial lowering of mean blood lead levels. However, because lead is a persistent metal, it is still present in the environment, water, soil, and dust (Patrick 2006).

In this regard, the work by Patterson in 1956 (Patterson 1956), who determined the age of the Earth by a uranium-lead isotopic data method, needs to be mentioned. Using a Canyon Diablo meteorite, Patterson was able to make an accurate measurement, calculating that the Earth was 4.55 billion years old. But in this study he discovered a disturbing and constant presence of lead in the atmosphere mainly due to tetraethyl lead used as an anti-knock gasoline. Thanks to the ongoing commitment of Patterson, tetraethyl lead, first in the United States and later in the rest of the world, was eliminated from gasoline. The presence of lead in the blood of human beings has considerably diminished; but in any case, the human beings today have about 625 times more lead in his body than people did 100–120 years ago.

In the majority of adults, chronic lead poisoning comes from exposures to work places and can occur in numerous work settings, such as manufacturing, lead smelting and refinement, or it may be caused by use of batteries, pigments, solder, ammunitions, paint, car radiators, cable and wires, and certain cosmetics (Brodkin et al. 2007). Diagnosis of lead toxicity has traditionally based on significantly elevated blood lead levels. These are an indicator of circulating lead that discloses variation in recent external lead exposure as well as of lead that has been mobilized by tissue stores (mostly bones). Lead levels in tibia and patella provide an indication of cumulative dose over decades (particularly cortical tissue in tibia) as well as the largest pool of lead in the body that is available for mobilization into blood. The latter phenomenon is heightened at times of high bone resorption (e.g. during pregnancy, aging, postmenopause) (Hu et al. 2007).

Inorganic lead is absorbed from the respiratory or gastrointestinal tract but not through the skin. Approximately 90 % of the total body burden is stored in bone and the remainder is in blood stream and soft tissue (Philip and Gerson 1994). Gastrointestinal absorption varies depending on nutritional status and age. Iron is believed to impair lead uptake in the gut, while iron deficiency is associated with increased blood lead concentrations in children. Lead exposure in pregnant animals usually occurs through the oral route. It known that absorption of this metal increase during pregnancy. Lead crosses the placenta and it accumulates in the fetus. Accumulation of lead occurs in the fetal brain owing to lack of blood-brain barrier (BBB). Lead also accumulates in the placenta in times of fetal stress (Gupta 2012).

Once absorbed, the circulating lead is bound to erythrocytes for approximately 30–35 %, while only 1 % of absorbed lead is found in plasma and serum and it is dispersed into the soft tissues of liver, renal cortex, aorta, brain, lungs, spleen, where it accumulates as $Pb_3(PO_4)_2$ in the following 4–6 weeks (Begovic et al. 2008). Lead is primarily excreted via the kidneys, while a small amount is excreted in feces and with sweat (Sinicropi et al. 2010b). The most common symptom of acute inorganic lead poisoning is gastrointestinal colic; chronic exposure to Pb^{2+} produce damage to hematopoietic, nervous, gastrointestinal and renal systems.

The effects of lead exposure are a health concern for all humans, but especially during early childhood because children are most at risk. Exposure to excessive amounts of inorganic lead during the toddler years may produce lasting adverse effects upon brain function. Maximal ingestion of lead occurs at an age when major changes are occurring in the density of brain synaptic connections.

Organolead compounds, as tetramethyllead and tetraethyllead are readily absorbed by inhalation and through the skin as well as by gastrointestinal tract. Tetraethyllead is metabolized to triethyllead, and this demethylated compound is excreted with the urine. Tetraethyllead and its metabolites are toxic especially for the brain. Toxicity appears with headache, restlessness, nervousness, and anxiety (Beattie et al. 1972); severe symptoms including convulsion, delirium, coma, abdominal pain and peripheral neuropathy. The neurotoxic effects of organolead compounds are associated with urinary lead concentrations higher than 30 mg/l (Macintyre 1994). In 2004, IARC classified lead and inorganic lead as probable human carcinogens (IARC group 2A), while organic lead remained unclassifiable.

3 Lead and Environment

Lead occurs naturally in the environment. It is rarely found in its elemental form but occurs in the Earth's crust primarily as the mineral galena (PbS), and to a lesser extent as anglesite ($PbSO_4$) and cerussite ($PbCO_3$). Lead minerals are found in association with zinc, copper, and iron sulfides as well as gold, silver, bismuth, and antimony minerals. It also occurs as a trace element in coal, oil, and wood.

Lead released from natural sources, such as volcanoes, windblown dust, and erosion, are minor compared with anthropogenic sources. In the air, lead is in the form of particles and is removed by rain or gravitational settling. The solubility of lead compounds in water is a function of pH, hardness, salinity, and the presence of humic material. Solubility is highest in soft and acidic water. The sink for lead is the soil and sediment. Because it is strongly adsorbed to soil, it generally is retained in the upper layers of soil and does not leach appreciably into the subsoil and groundwater.

Anthropogenic sources of lead include the mining and smelting of ore, manufacture of lead-containing products, combustion of coal and oil, and waste incineration. Many anthropogenic sources of lead, most notably leaded gasoline, lead-based paint, lead solder in food cans, lead-arsenate pesticides have been eliminated or strictly regulated due to lead's persistence and toxicity. Because lead does not degrade, these former uses leave their legacy as higher concentrations of lead in the environment.

Lead occurs naturally in the environment on account from human activities and it continues to be a significant public health problem in developing countries (Tong and McMichael 1999; Grant and Davis 1989), where there are considerable variations in the sources and pathways of exposure (Environmental Protection Agency, EPA 1986).

Exposure attributable to miscellaneous sources may be even more significant than universal exposure associated with leaded petrol, especially for people living in poverty (IPCS 1995, Environmental health criteria 165).

Exposure to lead from lead mining (Ajumobi et al. 2014), smelting of leads ores, as well as other ores (zinc, copper, iron, gold (Dooyema et al. 2012) and silver) in which lead is by-product or contaminant, battery factories and cottage industries is a significant environmental hazard in developing countries. Electrical utilities release into the atmosphere lead in flue gas from the burning fuels, such as coal, in which this element is a contaminant. As a result of human activity, environmental levels of lead increased more than 100-fold over the past three centuries.

The greatest increase occurred in the past century between the years 1950–2000 and reflected increasing worldwide use of tetraethyl lead and tetramethyl lead as gasoline additives to increase octane rating. Since gasoline additives have been banned, the level of lead in the atmosphere has dropped dramatically. Tetraethyl and tetramethyl lead, once added to gasoline, are no longer present in significant quantities in air. In fact, exposed to sunlight, they decompose rapidly to trialkyl and dialkyl lead compounds and to lead oxides by direct photolysis, and reacting with hydroxyl radicals and ozone. But it is necessary to emphasize that in the winter tetraethyl and tetramethyl lead have half-lives of up to several days since the atmospheric hydroxyl radicals concentration is lower than in summer (DeJonghe and Adams 1986).

In parallel with enforcement of the reduction of lead gasoline in Italy, Annibaldi and coauthors (2009) have studied the lead content of Adriatic seawater. In the years from 2000 to 2004, seawater was collected systematically at three sites along the coast line close to the city of Ancona. The results show that the lead content in seawater diminished from a median value of 0.25 nmol/L in 2000–2001 to 0.12 nmol/L in 2003–2004. This decrease has been correlated to the concurrent decrease of lead in gasoline in Italy with a reduction of lead emission in atmosphere.

Lead is also released into the air during burning coal and oil. In fact, in the last 15–20 years the total lead emission from electric steam increased due to the increased demand for electric power and an increased use of coal and natural gas as fuel sources to generate electricity. Burning these sources of energy recklessly, the level of CO_2 and thus global warming due to greenhouse effect have increased considerably.

Once small lead particles get into the atmosphere, they can travel long distances about 10 km from emission sources (Berndtsson 1993), before to fall by rain to land or into surface of rivers, lakes and sea. Sources of lead in dust and in soil include not only lead that falls to the ground from the air, but also weathering and chipping of lead-based paint from buildings and bridges. Higher levels of lead in soil are found near roadways (Nielsen 1984). Once lead falls onto soil, it sticks strongly to soil particles for many years in the upper layer of soil. Small amounts of lead may enter in rivers, lakes and sea when the soil particles are moved by rain water, or when lead is released by acidic water.

The fate of lead in soil is affected by the absorption at mineral interfaces and the formation of relatively stable organic-metal complexes or chelates with soil organic matter. This process is dependent on factor such as soil pH, soil type, organic matter content of soil, and cation exchange capacity (Reddy et al. 1995). Most lead is strongly retained in soil and very little is transported through runoff to surface of water. Clays, silts, iron and manganese oxides, and soil organic matter may bind lead electrostatically as cation exchange resin, as well as chemically for specific adsorption (Reed et al. 1995).

The amount of soluble lead in surface water depends upon the pH of the water and the concentration of dissolved salts. Equilibrium calculations show that at pH 5.4 the solubility of lead is about 30 μg/L in hard water and approximately 500 μg/L in soft water. Sulfate ions, if present in soft water, decrease the lead concentration through the formation of insoluble lead sulfate. The lead carbonate limits the amount of soluble lead considering also the partial pressure of CO_2, pH and temperature (Environmental Protection Agency, EPA 1986).

Plants and animals may bioconcentrate lead and the high lead concentrations are found in aquatic and terrestrial organisms. This occurs when these living beings have habitats near lead mining and smelting, areas affected by high automobile and truck traffic, sewage sludge and spoil disposable areas, sites where dredging have occurred, and in urban and industrialized areas (McGrath et al. 1994). Lead may be present on plant surfaces on account of atmosphere deposition; but its presence in internal plant tissues indicates biological uptake from the soil and leaf surface. Lead may be taken up in edible vegetables and fruits from the soil via the root system, by direct foliar uptake and translocation within the plant. As already mentioned, the amount of lead in soil, that is bioavailable to a vegetable plants, depends on factor such as cation exchange capacity, pH of soil, amount of organic matter present and type of fertilizer added to the soil (Holmgren et al. 1993).

Uptake of lead in animals may occur on account of inhalation of contaminated ambient air or ingestion of contaminated plants. However, lead is not biomagnified in aquatic or terrestrial food chains, as for other metals, for example mercury. In aquatic organisms, lead levels are usually highest in benthic organisms and algae, and lowest in upper trophic level predators as carnivorous fishes (Tulasi et al. 1992). Lead is toxic to all aquatic biotic component, and organisms higher up in the food chain may experience lead poisoning by ingestion of food contaminated with lead. Depuration is relatively rapid; in the case of rainbow trout exposed to tetramethyl lead the half-life values of depuration are about 35–45 h (Eisler 1988).

4 Lead Effect on Health

4.1 Effect on the Nervous System

Both central and peripheral nervous systems (CNS and PNS, respectively) appear to be the most sensitive targets for lead induced toxicity (Cory-Slechta 1996). While PNS is more affected in adults, the effect of lead on CNS is more pronounced in children (Bellinger 2004; Brent 2006). Once in the brain, lead-induced damage in the prefrontal cerebral cortex, hippocampus, and cerebellum can lead to a variety of neurological disorders, such as brain damage, mental retardation, behavioral problems, nerve damage, and possibly Alzheimer's disease, Parkinson's disease, and schizophrenia. Lead toxicity leads to encephalopathy with a progressive degeneration of certain regions of brain, the major symptoms including dullness, loss of memory, hallucinations, headache, poor attention span, and irritability.

At very high levels lead exposure, severe manifestations occur with delirium, convulsion and coma (Flora et al. 2006). Lead has negative effect especially on the developing nervous system of the fetuses and young children, which absorbs a higher fraction of this metal. The level of systematically circulating lead that has access to the brain of children is significantly higher as compared to adult subjects (Needleman 2004). At low levels exposure, children may appear inactive, hyperactive and irritable; moreover, IQ and concentration ability are significantly lowered. In the presence of greater lead levels, children show growth retardation, decreased intelligence, short-term memory and hearing loss. At last, at higher levels, children suffer permanent brain damage and even death (Cleveland et al. 2008). Besides, repercussions on the nervous system have been observed involving reduced motor activity, due to the loss of the insulating layer of myelin; in this way, a weakening of the nerve signal occurs, causing muscular weakness, especially of the exterior muscles, fatigue and lack of muscular coordination (Sanders et al. 2009).

4.2 Effect on the Hematopoietic System

Lead affects the hematopoietic system by inhibiting the synthesis of hemoglobin, acting directly on three key enzymes involved in the heme synthesis (Fig. 1) and, ultimately, leading to anemia. Lead affects the heme synthesis pathway by

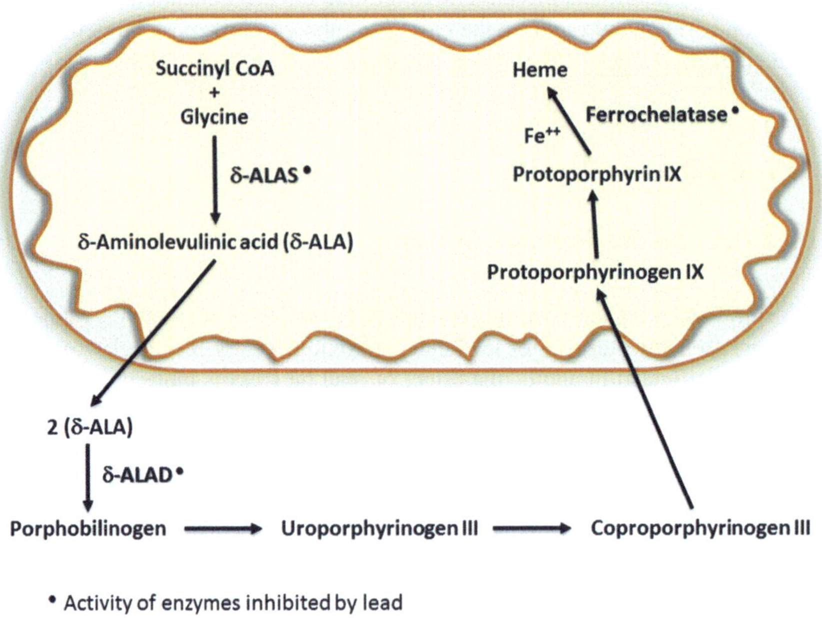

Fig. 1 Inhibition of lead in the heme biosynthetic pathway. The initial and final steps of heme synthesis take place in the mitochondria, while the intermediate steps develop in the cytoplasm. δ-ALAS: δ-aminolevulinic acid synthase; δ-ALAD: δ-aminolevulinic acid dehydratase; Succinyl CoA: Succinyl Coenzyme A

downregulating the three key enzymes in a dose dependent manner. The mitochondrial enzyme δ-aminolevulinic acid synthase (δ-ALAS) catalyzes the synthesis of δ-aminolevulinic acid (δ-ALA), starting from glycine and succinyl CoA (an intermediate of tricarboxylic acid cycle or Krebs cycle). Porphobilinogen is produced from two δ-ALA molecules, in the presence of the cytosolic enzyme δ-aminolevulinic acid dehydratase (δ-ALAD). Finally, the mitochondrial enzyme ferrochelatase catalyzes the insertion of a ferrous ion (Fe^{2+}) into protoporphyrin IX to form heme (Piomelli 2002).

δ-ALAD is a crucial enzyme in lead toxicity (Fig. 1); its inhibition decreases heme production and increases the quantity of δ-ALA that can be found in blood and urine of subjects with lead exposure (Bechara 1996). Heme synthesis does not decreases until the activity of δ-ALAD is inhibited by 80–90 %, which occurs at a blood lead concentration of about 55 μg/dl (Ahamed et al. 2005). It has also been shown that δ-ALA, when accumulating during lead exposure, autoxidizes with the resulting conversion of oxyhemoglobin to methemoglobin (Monteiro et al. 1986).

Ferrochelatase inhibition by lead allows the substitution of iron by zinc producing zinc protoporphyrin (ZPP). Thus, the concentration of ZPP increases and its presence can be used as an indicator of lead exposure level (Jangid et al. 2012). Inhibition of ferrochelatase results in increased excretion in urine of coproporphyrinogen and accumulation of protoporphyrin in erythrocytes.

Lead also reduces the life span of erythrocytes in the bloodstream. Erythrocytes bind about 98–99 % of the lead in the bloodstream and this metal has a destabilizing effect on cellular membranes. In red blood cells (RBC) lead causes a decrease of cell membrane fluidity and an increase of erythrocyte hemolysis rate and this is associated to anemia (Vij 2009). Hemolysis is the final result of ROS-generated lipid peroxidation in the RBC membranes and in all cellular membranes. Lead can also bind directly to phospholipids (particularly to phosphatidylcholine) in the RBC membranes, reducing their levels.

4.3 Effect on the Reproductive System

Lead interferes with the reproductive system. In a scientific study, semen quality from 100 workers occupationally exposed to variable quantities of lead was compared with the quality of about 150 volunteers with no exposure to lead (Telisman et al. 1990). Average level of lead in the blood of workers control was 10.7 μg/dl (range 6.7–20.8 μg/dl), while that of exposed workers was 37.1 μg/dl (range 11.7–104.0 μg/dl). The presence of higher amounts of lead in the blood of exposed workers compared to control workers (about 3.5 times) reduced the volume of ejaculation, semen density, total sperm number and motility, and increased the percentage of pathological spermatozoa (Goyer 1993). Other effects of high levels of blood lead include reduced libido, abnormal spermatogenesis, chromosomal damage, infertility and changes in serum testosterone. Moreover, during a study on men without occupational exposure to lead, it was demonstrated that high levels of lead in semen reduce the sperm count, contributing to its infertility (Wu et al. 2012).

It has been recognized that women with severe lead intoxication are more susceptible to prolonged and abnormal menstruations, infertility, miscarriage, stillbirth, premature membrane rupture, pregnancy hypertension and premature delivery (Flora et al. 2011). Besides, during pregnancy, direct influence of lead on the developmental stages of fetus has also been reported (Saleh et al. 2009). Moreover, the transfer of lead through the placenta into the mother milk makes the blood lead levels of the mothers and infants usually similar (Dart et al. 2004).

4.4 *Effect on the Kidney*

Lead can cause acute and chronic nephropathies. Lead is absorbed by the proximal tubular cells of the renal tubules, where it binds to specific lead-binding proteins. In presence of acute lead nephrotoxicity, these lead-protein complexes are observed as typical intracellular inclusion bodies. They do not secrete proteins in urine, but generates an abnormal excretion of glucose, phosphates and aminoacids, a combination known as Fanconi's syndrome.

On the other hand, chronic lead nephropathy is much more severe and causes irrevocable morphological and functional changes, such as glomerular and tubulointerstitial changes accompanied by hypertension, hyperuricemia and renal breakdown (Rastogi 2008).

Lead accumulates in the kidney mitochondria and causes both structural and functional alterations. These effects include mitochondrial swelling and inhibition of respiratory chain function and oxidative phosphorylation for ATP production. Consequently, energy-dependent processes, including tubular transport, are impaired.

4.5 *Effect on the Bone*

It is well recognized that lead has effect on bone metabolism. It accumulates in human body primarily in the bones (Silbergeld et al. 1993; Renner 2010). Lead is stored in two bone compartments: the exchangeable lead is present at the surface of bone and the non-exchangeable lead is located deeply in the cortical bone. Lead in bone can be mobilized during different physiological and pathological states. These conditions include endocrine status, age, osteoporosis, and maternal age during pregnancy and lactation (Silbergeld 1991).

5 Molecular Mechanism of Lead Toxicity: Oxidative Stress and Cation Action

Oxidative stress represents an imbalance between the production of free radicals and the cells ability to detoxify the extremely reactive intermediates or to repair the resulting damage (Flora et al. 2011). Oxidative stress occurs as a consequence of two different and related pathways: the generation of reactive oxygen species (ROS), like hydroperoxides (HO_2^-), singlet oxygen and hydrogen peroxide (H_2O_2); and the direct depletion of antioxidant reserves (Ercal et al. 2001; Flora 2002). In any biological system where ROS production increases, antioxidant reserves are depleted.

The tripeptide glutathione (GSH, γ-glutamylcysteinylglicine) is the most important antioxidant in cells; glutathione has a sulfhydryl group (-SH), present in cysteine residue, and is found in millimolar concentrations in mammalian tissues. Glutathione exists in two forms: reduced GSH and oxidized glutathione disulfide (GSSG) forms. Glutathione in the reduced state acts as an important antioxidant for quenching free radicals, donating reducing equivalent to ROS and makes them stable. In the redox reaction, GSH readily combines with another molecule of GSH and forms oxidized GSSG in the presence of the enzyme glutathione peroxidase (GPx). In turn, GSH can be restored from GSSG by the enzyme glutathione reductase (GR). Under normal conditions, 90 % of total glutathione exists in cells as reduced form GSH, and about 10 % as oxidized form GSSG; on the contrary, under oxidative stress, the GSSG concentration is much higher than that of GSH (Mates 2000; Flora et al. 2012). Lead covalently interacts with –SH groups of glutathione and antioxidant enzymes, inactivating them.

Lead inactivates not only GR, but also GPx, and glutathione-*S*-transferase, which further depress the glutathione levels (Hunaiti et al. 1995; Kasperczyk et al. 2004; Ahamed and Siddiqui 2007). Other important antioxidant enzymes inhibited by lead include superoxide dismutase (SOD) and catalase (CAT). Moreover, lead can also takes the place of the zinc ion, which acts as an important cofactor, in the catalytic site of various antioxidant enzymes, inhibiting them (Flora et al. 2007). Lead toxicity arises also on account of its ability to substitute other monovalent and divalent cations (Na^+, Ca^{2+}, Mg^{2+}, Fe^{2+}), affecting fundamental and important biological functions of the body (Lidsky and Schneider 2003). These fundamental cellular processes include cellular signaling, cell adhesion, apoptosis, enzyme regulation, and release and uptake of neurotransmitters (choline, dopamine and GABA) (Bressler et al. 1999; Garza et al. 2006). The ability of lead to pass through the BBB is due in large part to its ability to substitute calcium ions. At the molecular level, lead interferes with the regulatory action of calcium on cell functions and disrupts many intracellular biological activities. After replacing calcium ions, lead contributes to neurological deficits and becomes able to cross the BBB at an appreciable rate. After crossing the BBB, lead accumulates in astroglial cell, containing lead binding proteins; the immature astroglial cells are also damaged from lead, which may prevent the formation of myelin sheath (Bressler et al. 1999).

The developmental reorganization of synapses is in part mediated by protein kinases, and these enzymes are particularly sensitive to stimulation by lead. By inappropriately activating specific protein kinases, lead poisoning may disrupt the development of neural networks without producing overt pathological alterations. Protein kinases appear to regulate the development of brain capillaries and the expression of the BBB properties. Stimulation of protein kinase by lead may disrupt barrier development and alter the precise regulation of the neuronal environment that is required for normal brain function (Goldstein 1990).

Plausible mechanisms of inorganic lead carcinogenicity include direct DNA damage, clastogenicity, or inhibition of DNA synthesis or repair. Since lead may generate ROS, it may cause oxidative damage to DNA, it can replace zinc in several proteins that function as transcriptional regulators, including protamines. Lead further reduces the binding of these proteins to recognition elements in genomic DNA, which suggests an epigenetic involvement of lead in altered gene expression. These events may be of particular relevance in transplacental exposures (Silbergeld et al. 2000). It has also been demonstrated that the ingestion of lead acetate may induce significant stimulation in aspartate aminotransferase (AST) and alanine aminotransferase (ALT) activities; moreover, total soluble protein and albumin contents of plasma are significantly decreased, the cholinesterase activity is inhibited, while the activities of alkaline and acid phosphates and lactate dehydrogenase are stimulated. Pb^{2+} ingestion reduce the contents of hemoglobin and RBC count of intoxicated rat's blood and the plasma levels of triiodothyronine (T3), thyroxine (T4) and white blood cells (WBC) count decrease (Ibrahim et al. 2012).

6 The Role of Antioxidants in Protecting Lead-Induced Oxidative Stress

Preventive measures are preferred over the treatment regimens, considering that once lead enters the body it is almost impossible to remove it completely or to reverse the harmful effects.

The oxidative stress stimulated by lead is a state that involves free radicals generation with decreasing of the antioxidant reserves and at the same time hampering the ability of the body to annihilate the negative effects of free radicals (Fig. 2). Free radicals generate a series of chain reactions that induce lipid peroxidation with cell membrane disruption, oxidation of proteins and nucleic acids DNA and RNA with cancer formation. The phospholipids of cell membranes, including RBC membranes, being constituted by polyunsaturated fatty acids with two or more double bonds, are more susceptible to oxidative stress induced by lead. In this peroxidation process, lead could affect the activity of membrane enzymes, endo- and exocytosis and signal transducing processes (Adonaylo and Oteiza 1999).

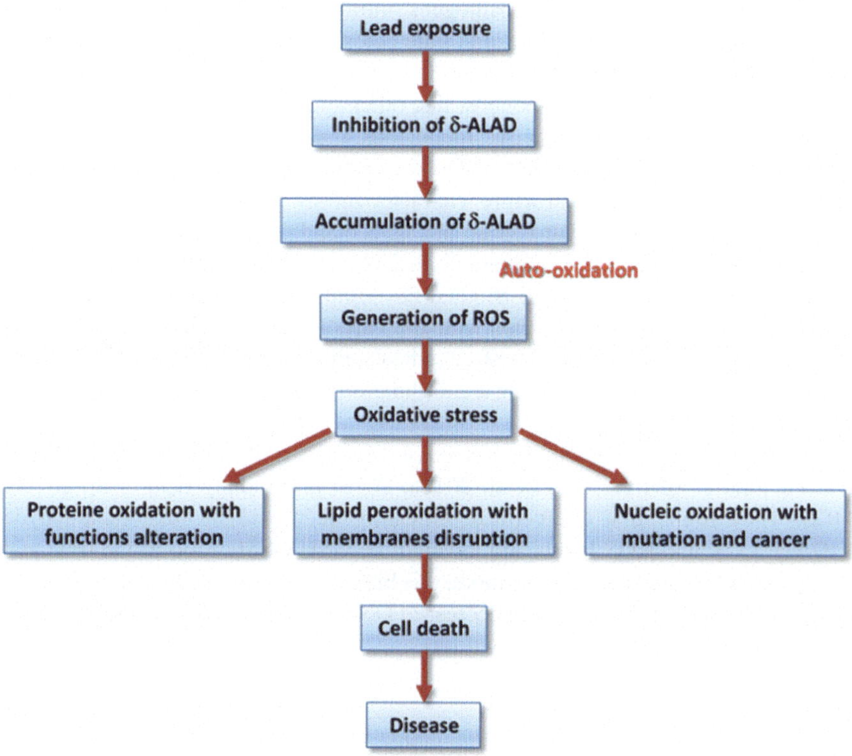

Fig. 2 Possible mechanism for lead-induced oxidative stress and cell death

Many authors suggested that administration of various antioxidants could prevent and cure the toxic effects of lead that causes the generation of free radicals in the body. They have the ability to scavenge ROS at molecular level and chelate lead ions, thereby reversing the toxic effects (Garcia and Gonzalez 2008).

The aforementioned enzymatic antioxidants, as SOD, CAT and GPx, are endogenously produced in the cells, whereas non-enzymatic antioxidants, like carotenoids, flavonoids, polyphenols, vitamins (vit B, vit C, vit E) are present in our daily food as fruits, vegetables, nuts, grains, meats and milk (Flora 2009). The non-enzymatic antioxidants are taken through the diet or in the form of supplements to maintain the homeostasis between free radicals and antioxidants.

6.1 *Flavonoids and Polyphenols*

Flavonoids are naturally polyphenolic compounds which represent the main constituents of fruits, vegetables, plant derived beverages (red wine and tea) and chocolate (Youdim et al. 2002). Flavonoids play an important role in plants, mainly protecting them against external pathogens, ultra-violet light, or heat. Flavonoids are responsible for the red, purple, and blue color of fruits and flowers, and play a role in pollination by attracting insects.

In human beings, these compounds, like other antioxidants, can cure or prevent oxidative stress by chelating active metal ions and also by terminating the free radical chain reaction (Terao 2009). The flavonoids ability to act as antioxidants depends on their molecular structure, characterized by two or more aromatic rings with at least one or more hydroxyl groups apiece and conjugated electrons giving also metal chelating properties (Heim et al. 2002; Wolfe and Liu 2008).

Quercetin is a ubiquitously distributed flavonoid present in fruit, vegetable and tea. The hydroxyl groups together with the carbonyl group donate electrons by undergoing resonance and stabilize free radicals, thus inhibiting lipid peroxidation. Quercetin chelates lead by forming a coordination bond between this metal and its ortho-phenolic groups (Fig. 3).

Liu and coworkers studied the protective mechanism of quercetin against lead-induced injuries of liver (Liu et al. 2010a, 2013) and kidney (Liu et al. 2010b, 2012). They found that quercetin significantly decreased the malondialdehyde (MDA), H_2O_2, and ROS levels and lowered the GSH/GSSG ratio in the liver and kidney of lead-treated rats. Furthermore, this bioflavonoid markedly restored Cu/Zn SOD, CAT and GPx activities and decreased DNA oxidative damage and apoptosis in the liver and kidney of lead-treated rats.

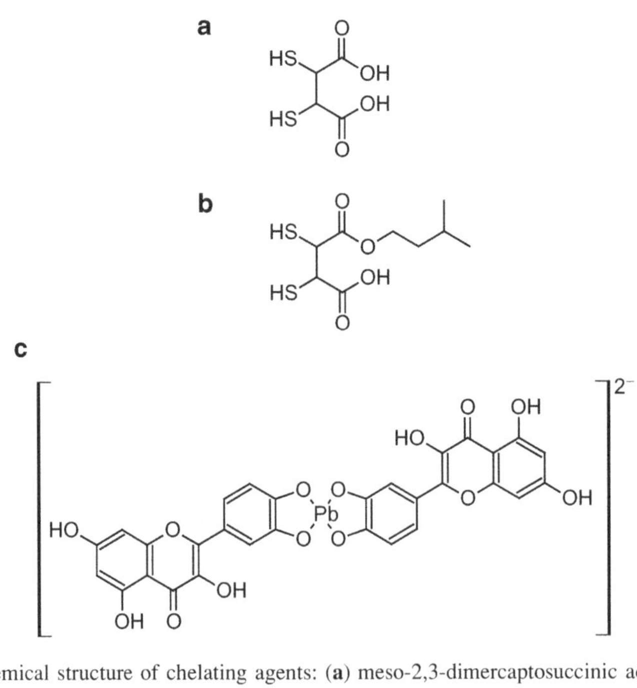

Fig. 3 Chemical structure of chelating agents: (**a**) meso-2,3-dimercaptosuccinic acid (DMSA); (**b**) monoisoamyl ester of DMSA (MiADMSA); (**c**) quercetin-Pb complex

Curcumin, a yellow polyphenolic compound, is the active component of tumeric, which possesses multiple activities, including antioxidant properties, radical scavenging and metal chelating in lead toxicity (Agarwal et al. 2010; Singh and Sankhla 2010). Shukla et al. (2003) reported the protective effect of curcumin against lead-induced neurotoxicity in rats. Exposure of rats to lead caused an increase in lipid peroxidation and a decrease in the level of reduced GSH, and in SOD and CAT activities in cerebellum, corpus striatum, hippocampus and frontal cortex as compared to controls. Treatment with curcumin caused in brain regions of the treated rats a significant decrease in lipid peroxidation and an important increase in reduced GSH level and SOD and CAT activities. Daniel et al. (2004) pointed out the chelation properties of curcumin, reducing lead level in rat brain. However, its extremely low aqueous-solubility and rapid intestinal and hepatic metabolism, which result in poor systemic bioavailability, restrict its oral use.

6.2 Vitamins

Vitamin B6 (pyridoxine) and vitamin B1 (thiamine) are the prostetic groups of the coenzymes pyridoxal 5′-phosphate and thiamine pyrophosphate: these vitamins are essential in the treatment of the deleterious effects of lead toxicity. Vitamin B6 indirectly acts as an antioxidant stimulating the synthesis of GSH and as a chelator agent (Ahamed and Siddiqui 2007). A diet rich in pyridoxine in lead-exposed rats improved δ-ALAD activity (Tandon et al. 1987); in addition, levels of lead in blood, kidney and liver were reduced after this particular diet. McGowan (1989) found that rats treated with lead and nourished with diet deficient in pyridine have significantly lower levels of GSH compared to lead-exposed rats and fed with normal levels of vitamin B6.

It has been reported that vitamin B1 exerts protective effect against lead toxicity. Senapati et al. (2000) reported the protective effect of thiamine hydrochloride on lead induced endogenous lipid peroxidation in liver and kidney of rats with an important decrement in the level of this metal.

Vitamin C is the most thoroughly studied vitamin in the prevention of lead induced oxidative stress. Its capability of quenching ROS together with metal chelation makes the ascorbic acid a potential detoxifying agent for lead (Tariq 2007).

Vitamin C (ascorbic acid) fortified with silymarin has been shown to reduce the toxic effect of acute lead poisoning on rat liver (Shalan et al. 2005). The combination of ascorbic acid with silymarin restored the activity of liver enzymes alanine aminotransferase (ALT), aspartate aminotransferase (AST) and gamma-glutamyl transpeptidase (γGT) after a lead induced damage. Combined treatment of lead-exposed animals with vitamin C and silymarin showed marked improvement of the biochemical, molecular and histopathological findings.

In rat liver supplementation of ascorbic acid and thiamine nullified the oxidative stress in a concentration dependent manner and protected DNA from damage induced by lead (Wang et al. 2007). In addition, Shan and coworkers (2009) reported that ascorbic acid and thiamine showed defensive results against the toxic effects of lead on testes of mice.

Vitamin E (α-tocopherol), a fat soluble vitamin, possess powerful antioxidative properties and in the membrane prevents lipid peroxidation by blocking the free radical chain reaction. Some authors (Sajitha et al. 2010) reported that α-tocopherol given to rats counteracted the harmful effect of lead by scavenging free radicals and preventing oxidative stress. δ-ALAD inhibition induced by lead in the erythrocytes is reversed by vitamin E treatment (Rendón-Ramirez et al. 2007).

Lead has been shown to decrease RBC membrane flexibility and to increase RBC fragility with risk for hemolysis (Levander et al. 1977). Vitamin E was shown to prevent RBC membrane damage on account of lead toxicity by lowering lipidic peroxide levels and increasing the activity of the enzymes SOD and catalase (Chaurasia and Kar 1997).

Vitamin E in combination with other antioxidant agents is more effective than its individual administration. Flora and coauthors (2003) studied the positive effects of naturally occurring antioxidants like vitamin C and vitamin E either alone or in combination with thiol chelators as meso-2,3-dimercaptosuccinic acid (DMSA) or monoisoamyl ester of DMSA (MiADMSA), on parameters indicative of oxidative stress in various organs (liver, kidney, brain and blood) of lead-exposed rats. The obtained results suggest that vitamin C and vitamin E administered during chelation with DMSA or MiADMSA were significantly beneficial in reducing oxidative stress. In fact, these thiol chelators (DMSA and MiADMSA) have two sulfhydryl (-SH) groups in the structure (Fig. 3) that can be useful in complexing lead and scavenging free radicals. This combination (vitamin and chelating agents) is more effective than the vitamin and the chelating agents alone, due to the fact that vitamin does not have lead chelating property but can consistently prevent oxidative stress.

Chelating agents form a complex with the toxic lead ion and these complexes, that show low toxicity, are easily eliminated from the body through the excretory system. An ideal chelating agent has to possess characteristics like great affinity for the toxic metal, that has to be chelated (in our case the lead), high water solubility, ability to cross cell membranes, possibility to oral administration and low metabolism. The uptake can be accomplished by passing through the phospholipid bilayers of the cellular membrane as an uncharged molecule or by utilizing a protein transport system embedded in the membrane.

6.3 Antioxidants Availability and Nanoparticles

There are several lines of evidence from *in vitro* and *in vivo* studies suggesting that antioxidants encapsulated in nanoparticles have a great potential in the prevention and treatment of various diseases. The major drawback in the usefulness of antioxidants appears to be due to poor solubility in aqueous solvents, poor absorption, poor bioavailability and rapid metabolism. To cancel these negative causes and to improve antioxidants usefulness numerous approaches have been undertaken, involving the use of nanoparticles, liposomes and phospholipid complex (Anand et al. 2007).

Lipid systems of nanoencapsulation enhance the antioxidants usefulness by improving their solubility and bioavailability and by preventing unwanted interactions with other food components. Liposome technology presents exciting opportunities in encapsulation and controlled release of antioxidants, as well as enhanced bioavailability and stability.

Mozafari and coauthors (2006, 2008) reported the use of liposomes as carrier vehicles of nutrients, enzymes, drugs, and food antimicrobials, because of their small size, biodegradability, hydrophobic and hydrophilic characters, and low toxicity. Results from several studies demonstrate that liposomes have the potential to enhance drug penetration, improve therapeutic effectiveness, and reduce serious side effects. Liposomes are tiny vesicles of spherical shape (usually between 100 and 1000 nm in diameter) containing water, artificially prepared and composed by phospholipids bilayers as the cellular membranes. Liposomes are prepared by sonication and are composed of phospholipids enriched of phosphatidylcholine and phosphatidylethanolamine. Phospholipids are amphiphilic with long hydrocarbon tails of the molecule being hydrophobic, while its polar head is hydrophilic.

The phospholipids have hydrophilic heads pointing outside, while the hydrophobic tails of both lipid bilayers point inside interacting with each other. Also the inside of the liposomes is water soluble and can contain soluble drugs, enzymes, nutrients and biomolecules. The outer phospholipids membranes can be covalently modified with charged molecules; in this way the liposomes may be more easily conveyed at the target cells.

Excellent results were obtained encapsulating curcumin in liposomes. Although curcumin, encapsulated into liposomes, is more bioactive and bioavailable, *in vivo* studies are yet necessary to confirm these properties (Gandhi et al. 2011).

7 Summary

Humans have used lead since early history for over 4000–5000 years for the most different applications. So lead poisoning has been known to men, but the situation worsened in the 18th century with the industrial revolution. In the last century the levels of lead in the human body increased because levels in the atmosphere increased due to the presence of this metal in the exhaust gases of cars powered by gasoline added with tetraethyl lead. Nowadays, the level of lead in our body is going down because it was banned from gasoline, however it is still present in many manufactured products and therefore produces deleterious toxic effects.

Lead released to the atmosphere partitions to surface water, soil and sediment. Organolead compounds (tetraethyl and tetramethyl lead) are transformed in the atmosphere by photodegradation and reaction with ozone to alkyl lead, oxides and carbonates; instead, in surface water organolead compounds are transformed by photolysis and hydrolysis. Anyhow, properties such as pH, oxygen content and salinity are necessary to fully understand the chemical transformation and the environmental fate of lead in soil and water.

It is important to stress that lead exposure induces generation of free radicals and ROS, resulting in oxidative damage to various biomolecules like nucleic acids (DNA and RNA), proteins and enzymes, and membranes based phospholipids, impairing at the same time the antioxidant defense system. In human beings, lead has no biological functions and once it enters the body, it causes severe irreversible health effects and it affects important systems as nervous, hematopoietic, reproductive and renal and so on.

Antioxidants, specifically vitamins (B_1, B_6, C and E) have been shown to lower ROS generated cellular damage. They have the ability to scavenge ROS and chelate lead ions, reversing the toxic effects. These antioxidants were also reported to provide an elevated therapeutic impact when administered with thiol chelators DSMSA and MiADMSA. These beneficial effects were accompanied by more pronounced urinary lead elimination and tissue lead depletion.

The biggest drawback in the case of antioxidants is their poor bioavailability due to low solubility and rapid clearance. Novel approaches to overcome the problem of low bioavailability of these antioxidants include the formulation of nanoparticles, liposomes and micelles. Compared to conventional methods, this approach reduces the dosage of antioxidants to maintain the therapeutic level in the body. Further experiments are needed to show the effects of nutrients on cells of animals and men that undergo the lead exposure.

Acknowledgements The authors gratefully acknowledge the financial support provided by Ministero dell'Istruzione, dell'Università e della Ricerca (MIUR, Italia).

Conflict of Interest The authors (AC, AC, GL, MSS, GG) declare no conflict of interest.

References

Adonaylo VN, Oteiza PI (1999) Pb^{2+} promotes lipid peroxidation and alteration in membrane physical properties. Toxicology 132:19–32

Agarwal R, Goel SK, Behari JR (2010) Detoxification and antioxidant effects of curcumin in rats experimentally exposed to mercury. J Appl Toxicol 30:457–468

Ahamed M, Siddiqui MK (2007) Environmental lead toxicity and nutritional factors. Clin Nutr 26:400–408

Ahamed M, Verma S, Kumar A, Siddiqui MK (2005) Environmental exposure to lead and its correlation with biochemical indices in children. Sci Total Environ 346:48–55

Ajumobi OO, Tsofo A, Yango M, Aworrh MK, Anagbogu IN, Mohammed A, Umar-Tsafe N, Mohammed S, Abdllahi M, Davis L, Idris S, Poggensee G, Nguku P, Gitta S, Nsubuga P (2014) High concentration of blood lead levels among young children in Bagega community, Zamfara - Nigeria and the potential risk factor. Pan Afr Med J 18(Suppl 1):14

Anand P, Kunnumakkara AB, Newman RA, Aggarwal BB (2007) Bioavailability of curcumin: problems and promises. Mol Pharm 4:807–818

Annibaldi A, Truzzi C, Illuminati S, Scarponi G (2009) Recent sudden decrease of lead in Adriatic coastal seawater during the years 2000–2004 in parallel with the phasing out of leaded gasoline in Italy. Mar Chem 113:238–249

Arvig E, Grandjean P, Beckmann J (1980) Neurotoxic effects of heavy lead exposure determined with psychological tests. Toxicol Lett 5:399–404

Baghurst PA, McMichael AJ, Wigg NR, Vimpani GV, Robertson EF, Roberts RJ, Tong SL (1992) Environmental exposure to lead and children's intelligence at the age of seven years. The Port Pirie Cohort Study. N Engl J Med 327:1279–1284

Baker EL (1982) Neurologic and behavioral disorders. In: Levy BS, Wegman DH (eds) Occupational health: recognition and prevention of work-related disease. Little, Brown and Co, Boston, MA

Baker EL, Feldman RG, White RA, Harley JP, Niles CA, Dinse GE, Berkey CS (1984) Occupational lead neurotoxicity: a behavioural and electrophysiological evaluation. Study design and year one results. Br J Ind Med 41:352–361

Basha R, Reddy GR (2010) Developmental exposure to lead and late life abnormalities of nervous system. Indian J Exp Biol 48:636–641

Beattie AD, Moore MR, Goldberg A (1972) Tetraethyl-lead poisoning. Lancet 300:12–15

Bechara EJ (1996) Oxidative stress in acute intermittent porphyria and lead poisoning may be triggered by 5-aminolevulinic acid. Braz J Med Biol Res 29:841–851

Begovic V, Nozic D, Kupresanin S, Tarabar D (2008) Extreme gastric dilation caused by chronic lead poisoning: a case report. World J Gastroenterol 14:2599–2601

Bellinger DC (2004) Lead. Pediatrics 113:1016–1022

Bellinger DC (2005) Teratogen update: lead and pregnancy. Birth Def Res A Clin Mol Teratol 73:409–420

Berndtsson R (1993) Small-scale spatial patterns of bulk atmospheric deposition. J Environ Qual 22:349–360

Brent JA (2006) Review of: "Medical Toxicology". Clin Toxicol 44:355

Bressler J, Kim KA, Chakraborti T, Goldstein G (1999) Molecular mechanisms of lead neurotoxicity. Neurochem Res 24:595–600

Brodkin E, Copes R, Mattman A, Kennedy J, Kling R, Yassi A (2007) Lead and mercury exposures: interpretation and action. Can Med Assoc J 17:59–63

Carocci A, Rovito N, Sinicropi MS, Genchi G (2014) Mercury toxicity and neurodegenerative effects. Rev Environ Contam Toxicol 229:1–18

Chaurasia SS, Kar A (1997) Protective effects of vitamin E against lead-induced deterioration of membrane associated type-1 iodothyronine 5′-monodeiodinase (5′D-I) activity in male mice. Toxicology 124:203–209

Cleveland LM, Minter ML, Cobb KA, Scott AA, German VF (2008) Lead hazards for pregnant women and children: Part 1: immigrants and the poor shoulder most of the burden of lead exposure in this country. Part 1 of a two-part article details how exposure happens, whom it affects, and the harm it can do. Am J Nurs 108:40–49

Cory-Slechta DA (1996) Legacy of lead exposure: consequences for the central nervous system. Otolaryngol Head Neck Surg 114:224–226

Daniel S, Limson JL, Dairam A, Watkins GM, Daya S (2004) Through metal binding, curcumin protects against lead- and cadmium-induced lipid peroxidation in rat brain homogenates and against lead-induced tissue damage in rat brain. J Inorg Biochem 98:266–275

Dart RC, Hurlbut KM, Boyer-Hassen LV (2004) Lead. In: Dart RC (ed) Medical toxicology, 3rd edn. Lippincott Williams & Wilkins, Philadelphia, PA, pp 1423–1431

DeJonghe WRA, Adams FC (1986) Biogeochemical cycling of organic lead compounds. Adv Environ Sci Technol 17:561–594

Dietert RR, Piepenbrink MS (2006) Lead and immune function. Crit Rev Toxicol 36:359–385

Dooyema CA, Neri A, Lo YC, Durant J, Dargan PI, Swarthout T, Biya O, Gidado SO, Haladu S, Sani-Gwarzo N, Nguku PM, Akpan H, Idris S, Bashir AM, Brown MJ (2012) Outbreak of fatal childhood lead poisoning related to artisanal gold mining in northwestern Nigeria, 2010. Environ Health Perspect 120:601–607

Eisler R (1988) Lead hazards to fish, wildlife, and invertebrates: a synoptic review, vol 85 (1.14), Biol report. U.S. Department of the Interior, Fish and Wildlife Service, Laurel, MD

EPA (1986) Air quality criteria for lead. U.S. Environmental Protection Agency, Office of Research and Development, Office of Health and Environmental Assessment, Environmental Criteria and Assessment Office, Research Triangle Park, NC, EPA600883028F

Ercal N, Gurer-Orhan H, Aykin-Burns N (2001) Toxic metals and oxidative stress. Part 1. Mechanisms involved in metal-induced oxidative damage. Curr Top Med Chem 1:529–539

Feldman RG, Ricks NL, Baker EL (1980) Neurophysiological effects of industrial toxins: a review. Am J Ind Med 1:211–227

Flora SJS (2002) Nutritional components modify metal absorption, toxic response and chelation therapy. J Nutr Environ Med 12:53–67

Flora SJS (2009) Structural, chemical and biological aspects of antioxidants for strategies against metal and metalloid exposure. Oxid Med Cell Longev 2:191–206

Flora SJ, Pande M, Mehta A (2003) Beneficial effect of combined administration of some naturally occurring antioxidants (vitamins) and thiol chelators in the treatment of chronic lead intoxication. Chem Biol Interact 145:267–280

Flora SJS, Flora GJS, Saxena G (2006) Environmental occurrence, health effects and management of lead poisoning. In: Cascas SB, Sordo J (eds) Lead: chemistry, analytical aspects, environmental impacts and health effects. Elsevier Publication, Netherlands, pp 158–228

Flora SJ, Flora G, Saxena G, Mishra M (2007) Arsenic and lead induced free radical generation and their reversibility following chelation. Cell Mol Biol (Noisy-le-Grand) 53:26–47

Flora SJS, Pachauri V, Saxena G (2011) Arsenic, cadmium and lead. In: Gupta RC (ed) Reproductive and developmental toxicology, vol 33. Academic, New York, NY, pp 415–438

Flora G, Gupta D, Tiwari A (2012) Toxicity of lead: a review with recent updates. Interdiscip Toxicol 5:47–58

Gandhi P, Khan Z, Chakraverty N (2011) Soluble curcumin: a promising oral supplement for health management. J Appl Pharm Sci 1:01–07

Garcia MTA, Gonzalez ELM (2008) Toxic effects of perinatal lead exposure on the brain of rats: involvement of oxidative stress and the beneficial role of antioxidants. Food Chem Toxicol 46:2089–2095

Garza A, Vega R, Soto E (2006) Cellular mechanisms of lead neurotoxicity. Med Sci Monit 12: RA57–RA65

Goldstein GW (1990) Lead poisoning and brain cell function. Environ Health Perspect 89:91–94

Gonick HC (2002) Lead, renal disease and hypertension. Am J Kidney Dis 40:202–204

Goyer R (1993) Lead toxicity: current concerns. Environ Health Perspect 100:177–187

Grant LD, Davis JM (1989) Effects of low-level lead exposure on paediatric neurobehavioural development: current findings and future directions. In: Smith MA, Grant LD, Sors AI (eds) Lead exposure and child development: an international assessment. MTP Press, Lancaster, pp 49–118

Gulson BL (2000) Revision of estimates of skeletal contribution to blood during pregnancy and postpartum period. J Lab Clin Med 136:250–251

Gupta RC (2012) Veterinary toxicology: basic and clinical principles, 2nd edn. Elsevier, London

Heim KE, Tagliaferro AR, Bobliya DJ (2002) Flavonoid antioxidants: chemistry, metabolism and structure activity relationships. J Nutr Biochem 13:572–584

Holmgren GGS, Meyer MW, Chaney RL, Daniels RB (1993) Cadmium, lead, cooper, and nickel in agricultural soils of the United States of America. J Environ Qual 22:335–348

Hu H, Rabinowitz M, Smith D (1998) Bone lead as a biological marker in epidemiologic studies of chronic toxicity: conceptual paradigms. Environ Health Perspect 106:1–8

Hu H, Shih R, Rothenberg S, Schwartz BS (2007) The epidemiology of lead toxicity in adults: measuring dose and consideration of other methodologic issues. Environ Health Perspect 115:455–462

Hunaiti A, Soud M, Khalil A (1995) Lead concentration and the level of glutathione, glutathione S-transferase, reductase and peroxidase in the blood of some occupational workers from Irbid City, Jordan. Sci Total Environ 170:95–100

Ibrahim NM, Eweis EA, El-Beltagi HS, Abdel-Mobdy YE (2012) Effect of lead acetate toxicity on experimental male albino rat. Asian Pac J Trop Biomed 2:41–46

International Agency for Research on Cancer (2006) Inorganic and organic lead compounds, vol 87, IARC monographs on the evaluation of carcinogenic risks to humans. International Agency for Research on Cancer (IARC), Lyon

IPCS (International Programme on Chemical Safety) (1995) Inorganic lead. Environmental health criteria 165. WHO, Geneva

Jangid AP, John PJ, Yadav D, Mishra S, Sharma P (2012) Impact of chronic lead exposure on selected biological markers. Indian J Clin Biochem 27:83–89

Kasperczyk S, Kasperczyk A, Ostalowska A, Dziwisz M, Birkner E (2004) Activity of glutathione peroxidase, glutathione reductase, and lipid peroxidation in erythrocytes in workers exposed to lead. Biol Trace Elem Res 102:61–72

Levander OA, Morris VC, Ferretti RJ (1977) Filterability of erythrocytes from vitamin E-deficient lead poisoned rats. J Nutr 107:363–372

Lidsky TI, Schneider JS (2003) Lead neurotoxicity in children: basic mechanisms and clinical correlates. Brain 126:5–19

Liu CM, Zheng YL, Lu J, Zhang ZF, Fan SH, Wu DM, Ma JQ (2010a) Quercetin protects rat liver against lead-induced oxidative stress and apoptosis. Environ Toxicol Pharmacol 29:158–166

Liu CM, Ma JQ, Sun YZ (2010b) Quercetin protects the rat kidney against oxidative stress-mediated DNA damage and apoptosis induced by lead. Environ Toxicol Pharmacol 30:264–271

Liu CM, Sun YZ, Sun JM, Ma JQ, Cheng C (2012) Protective role of quercetin against lead-induced inflammatory response in rat kidney through the ROS-mediated MAPKs and NF-κB pathway. Biochim Biophys Acta 1820:1693–1703

Liu CM, Zheng GH, Ming QL, Sun JM, Cheng C (2013) Protective effect of quercetin on lead-induced oxidative stress and endoplasmic reticulum stress in rat liver via the IRE1/JNK and PI3K/Akt pathway. Free Radic Res 47:192–201

Macintyre JE (1994) Dictionary of organometallic compounds. CRC Press, Boca Raton, FL. ISBN ISBN 9780412430602

Mates JM (2000) Effects of antioxidant enzymes in the molecular control of reactive oxygen species toxicology. Toxicology 153:83–104

McGowan C (1989) Influence of vitamin B6 status on aspects of lead poisoning in rats. Toxicol Lett 47:87–93

McGrath SP, Chang AC, Page AL, Witter E (1994) Land application of sewage sludge: scientific perspectives of heavy metal loading limits in Europe and the United States. Environ Rev 2:108–118

Monteiro HP, Abdalla DSP, Faljoni-Alario A, Bechara EJH (1986) Generation of active oxygen species during coupled autoxidation of oxyhemoglobin and delta-aminolevulinic acid. Biochim Biophys Acta 881:100–106

Mozafari MR, Flanagan J, Matia-Merino L, Awati A, Omri A, Suntres Z, Singh H (2006) Recent trends in the lipid-based nanoencapsulation of antioxidants and their role in foods. J Sci Food Agric 86:2038–2045

Mozafari MR, Johnson C, Hatziantoniou S, Demetzos C (2008) Nanoliposomes and their applications in food nanotechnology. J Liposome Res 18:309–327

Needleman H (2004) Lead poisoning. Annu Rev Med 55:209–222

Nielsen T (1984) Atmospheric occurrence of organolead compounds. In: Grandjean P (ed) Biological effects of organolead compounds. CRC Press, Boca Raton, FL, pp 43–62

Patrick L (2006) Lead toxicity, a review of the literature. Part 1: Exposure, evaluation, and treatment. Altern Med Rev 11:2–22

Patterson C (1956) Age of meteorites and the earth. Geochim Cosmochim Acta 10:230–237

Pearce JM (2007) Burton's line in lead poisoning. Eur Neurol 57:118–119

Philip AT, Gerson B (1994) Lead poisoning--Part I. Incidence, etiology, and toxicokinetics. Clin Lab Med 14:423–444

Piomelli S (2002) Childhood lead poisoning. Pediatr Clin North Am 49:1285–1304

Rastogi SK (2008) Renal effects of environmental and occupational lead exposure. Ind J Occup Environ Med 12:103–106

Reddy KJ, Wang L, Gloss SP (1995) Solubility and mobility of copper, zinc and lead in acidic environments. Plant Soil 171:53–58

Reed BE, Moore RE, Cline SR (1995) Soil flushing of a sandy loam contaminated with Pb(II), $PbSO_4$ (s), $PbCO_3$ (3) or Pb-naphthalene: column results. J Soil Contam 4:243–267

Rendón-Ramirez A, Cerbón-Solórzano J, Maldonado-Vega M, Quintanar-Escorza MA, Calderón-Salinas JV (2007) Vitamin-E reduces the oxidative damage on delta-aminolevulinic dehydratase induced by lead intoxication in rat erythrocytes. Toxicol In Vitro 21:1121–1126

Renner R (2010) Exposure on tap: drinking water as an overlooked source of lead. Environ Health Perspect 118:A68–A74

Sajitha GR, Jose R, Andrews A, Ajantha KG, Augustine P, Augusti KT (2010) Garlic oil and vitamin E prevent the adverse effects of lead acetate and ethanol separately as well as in combination in the drinking water of rats. Indian J Clin Biochem 25:280–288

Saleh HA, El-Aziz GA, El-Fark MM, El-Gohary M (2009) Effect of maternal lead exposure on craniofacial ossification in rat fetuses and the role of antioxidant therapy. Anat Histol Embryol 38:392–399

Sanders T, Liu Y, Buchner V, Tchounwou PB (2009) Neurotoxic effects and biomarkers of lead exposure: a review. Res Environ Health 24:15–45

Senapati SK, Dey S, Dwivedi SK, Patra RC, Swarup D (2000) Effect of thiamine hydrochloride on lead induced lipid peroxidation in rat liver and kidney. Vet Hum Toxicol 42:236–237

Shalan MG, Mostafa MS, Hassouna MM, El-Nabi SE, El-Refaie A (2005) Amelioration of lead toxicity on rat liver with vitamin C and silymarin supplements. Toxicology 206:1–15

Shan G, Tang T, Zhang X (2009) The protective effect of ascorbic acid and thiamine supplementation against damage caused by lead in the testes of mice. J Huazhong Univ Sci Technolog Med Sci 29:68–72

Shukla PK, Khanna VK, Khan MY, Srimal RC (2003) Protective effect of curcumin against lead neurotoxicity in rat. Hum Exp Toxicol 22:653–658

Silbergeld EK (1991) Lead in bone: implications for toxicology during pregnancy and lactation. Envirom Health Perspect 91:63–70

Silbergeld EK (2003) Facilitative mechanisms of lead as a carcinogen. Mutat Res 533:121–133

Silbergeld EK, Schwart J, Mahaffey K (1988) Lead and osteoporosis: mobilization of lead from bone in postmenopausal women. Environ Res 47:79–94

Silbergeld EK, Sauk J, Somerman M, Todd A, McNeill F, Fowler B, Fontaine A, van Buren J (1993) Lead in bone: storage site, exposure source, and target organ. Neurotoxicology 14:225–236

Silbergeld EK, Waalkes M, Rice JM (2000) Lead as a carcinogen: experimental evidence and mechanisms of action. Am J Ind Med 38:316–323

Singh P, Sankhla V (2010) In situ protective effect of curcumin on cadmium chloride induced genotoxicity in bone marrow chromosomes of Swiss albino mice. J Cell Mol Biol 8:57–64

Sinicropi MS, Amantea D, Caruso A, Saturnino C (2010a) Chemical and biological properties of toxic metals and use of chelating agents for the pharmacological treatment of metal poisoning. Arch Toxicol 84:501–520

Sinicropi MS, Caruso A, Capasso A, Palladino C, Panno A, Saturnino C (2010b) Heavy metals: toxicity and carcinogenicity. Pharmacology 2:329–333

Somervaille LJ, Chettle DR, Scott MC, Tennant DR, McKiernan MJ, Skilbeck A, Trethowan WN (1988) In vivo tibia lead measurements as an index of cumulative exposure in occupationally exposed subjects. Br J Ind Med 45:174–181

Steenland K, Boffetta P (2000) Lead and cancer in humans: where are we now? Am J Ind Med 38:295–299

Stollery BT, Banks HA, Broadbent DE, Lee WR (1989) Cognitive functioning in lead workers. Br J Ind Med 46:698–707

Stollery BT, Broadbent DE, Banks HA, Lee WR (1991) Short term prospective cognitive functioning in lead workers. Br J Ind Med 48:739–749

Tandon SK, Flora SJ, Singh S (1987) Influence of pyridoxine (vitamin B6) on lead intoxication in rats. Ind Health 25:93–96

Tariq SA (2007) Role of ascorbic acid in scavenging free radicals and lead toxicity from biosystems. Mol Biotechnol 37:62–65

Telisman S, Cvitkovic P, Gavella M, Pongracic J (1990) Semen quality in men with respect to blood lead and cadmium levels. In International symposium on Lead and cadmium toxicology. Peking, People's Republic of China, pp 29–32

Terao J (2009) Dietary flavonoids as antioxidants. Forum Nutr 61:87–94

Tong S, McMichael AJ (1999) The magnitude, persistence and public health significance of cognitive effects of environmental lead exposure in childhood. J Environ Med 1:103–110

Tulasi SJ, Reddy PUM, Rao JV (1992) Accumulation of lead and effects on total lipids and lipid derivatives in the freshwater fish *Anabas testudineus* (Bioch). Ecotoxicol Environ Saf 23:33–38

van Wijngaarden E, Dosemeci M (2006) Brain cancer mortality and potential occupational exposure to lead: findings from the National Longitudinal Mortality Study, 1979–1989. Int J Cancer 119:1136–1144

Vaziri ND (2002) Pathogenesis of lead-induced hypertension: role of oxidative stress. J Hypertens 20:S15–S20

Vij AG (2009) Hemopoietic, hemostatic and mutagenic effects of lead and possible prevention by zinc and vitamin C. Al Ameen J Med Sci 2:27–36

Wang C, Liang J, Zhang C, Bi Y, Shi X, Shi Q (2007) Effect of ascorbic acid and thiamine supplementation at different concentrations on lead toxicity in liver. Ann Occup Hyg 51:563–569

Wolfe KL, Liu RH (2008) Structure–activity relationships of flavonoids in the cellular antioxidant activity assay. J Agric Food Chem 56:8404–8411

Wu HM, Lin-Tan DT, Wang ML, Huang HY, Lee CL, Wang HS, Soong YK, Lin JL (2012) Lead level in seminal plasma may affect semen quality for men without occupational exposure to lead. Reprod Biol Endocrinol 10:91–95

Youdim KA, Spencer JP, Schroeter H, Rice-Evans C (2002) Dietary flavonoids as potential neuroprotectants. Biol Chem 383:503–519

Persistent Organic Pollutants and Concern Over the Link with Insulin Resistance Related Metabolic Diseases

Sara Mostafalou

Contents

Abbreviations

AhR	Aryl hydrocarbon receptor
BFR	Brominated flame retardants
BMI	Body mass index
CARDIA	Coronary Artery Risk Development in Young Adults
CHL	Chlordane
COX	Cyclooxygenase
CRP	C-reactive protein
DDE	*p,p′*-dichlorodiphenyltrichloroethane
DDT	Dichlorodiphenyltrichloroethane
HCB	Hexachlorobenzene
HCH	Hexachlorocyclohexane
HOMA	Homeostatic model assessment

S. Mostafalou, Pharm.D., Ph.D. (✉)
Department of Pharmacology and Toxicology, School of Pharmacy, Ardabil University of Medical Sciences, 5618953141 Ardabil, Iran
e-mail: s.mostafalou@arums.ac.ir; s.mostafalou@gmail.com

P. de Voogt (ed.), *Reviews of Environmental Contamination and Toxicology, Volume 238*, Reviews of Environmental Contamination and Toxicology 238,
DOI 10.1007/398_2015_5001

HpCDD	Heptachlorodibenzo-p-dioxin
IL-6	Interleukin 6
IR	Insulin resistance
NF-κB	Nuclear factor kappa-light-chain-enhancer of activated B cells
OCDD	Octachlorodibenzo-p-dioxin
OCP	Organochlorine pesticide
PAH	Polycyclic aromatic hydrocarbons
PBB	Polybrominated biphenyls
PBDE	Polybrominated diphenyl ethers
PCB	Polychlorinated biphenyls
PCDD	Polychlorodibenzo-*p*-dioxin
PCDF	Polychlorodibenzofuran
PFNA	Perfluorononanoic acid
PIVUS	Prospective Investigation of the Vasculature in Uppsala Seniors
POP	Persistent organic pollutants
TBT	Tributyltin
TCDD	Tetrachlorodibenzo-p-dioxin
TNF-α	Tumor necrosis factor alpha

1 Introduction

Persistent organic pollutants (POPs) refer to organic compounds tending to persist in the environment due to their high resistance to biological, chemical, and photolytic degradation, and can have a long range transport due to their semi-volatility. In 1995, they were introduced as "dirty dozen" chemicals including aldrin, chlordane, DDT, dieldrin, endrin, heptachlor, hexachlorobenzene, mirex, polychlorinated biphenyls, polychlorinated dibenzo-p-dioxins, polychlorinated dibenzofurans, and toxaphene. Then after some other compounds like polycyclic aromatic hydrocarbons (PAHs), brominated flame-retardants, and tributyltin (TBT) were also added to that list. These chemicals are highly lipid soluble which can lead to their bioaccumulation in fat tissues and biomagnifications in food chain. The Stockholm Convention on Persistent Organic Pollutants signed in 2001 and implemented in 2004, the production and use of known POPs were eliminated by parties. Due to this international treaty, the environmental levels of POPs was effectively decreased during the last decade, but a wide scientific agreement has been reached that different concentrations of POPs can be found in everyone's body, yet. On the other side, increasing evidences are being provided day by day regarding POPs health hazards focusing on their disruptive effects on endocrine, reproductive, and immune systems, and carcinogenic properties (Ritter et al. 2007).

The main route of human exposure to POPs is dietary intakes particularly foods of animal origin like fatty fish, meat, poultry and dairy products. Upon absorption in human body, these chemicals are well distributed toward fatty tissues leading to their accumulation in fat and secretion into the milk (Halldorsson 2012; Jurjanz et al. 2008).

Recently, POPs have attracted much attention in association with metabolic disorders so these compounds have been proposed as environmental risk factors for type 2 diabetes (Mostafalou and Abdollahi 2012). Although, obesity is considered as the main promoter of type 2 diabetes, there has been much doubt on the role of POPs in developing metabolic diseases riding on insulin resistance (Lee et al. 2006a, b; Ngwa et al. 2015; Ruzzin 2012; Wang et al. 2010a).

Insulin resistance (IR) as a pathologic mal-adaption of insulin signaling pathway has been much thought to be caused by exposure to POPs (Tillett 2010; Mostafalou and Abdollahi 2013). Some epidemiologic studies have reported high levels of POPs in the serum and urine of IR patients (Kim et al. 2012a, b; Lee et al. 2007a, 2011). There are also some experimental evidences on the link between exposure to these compounds and metabolic biomarkers of IR in target tissues of insulin (Baker et al. 2013; Gray et al. 2013; Ibrahim et al. 2011; Mostafalou et al. 2012). One of the main risk factors of IR is obesity by which the normal response of adipocytes to insulin is disrupted. Furthermore, obesity facilitates the distribution of lipophilic compounds from systemic circulation to accumulate in fat mass. High potential of POPs to deposit in fat tissues has raised the assumptions that these chemicals can cause development of IR through not only disruption of insulin signaling cascade but also amplification of the other risk factor, obesity (Lee 2012).

One of the main proposed mechanisms for POPs-induced metabolic disorders is chronic inflammation which can disrupt the normal metabolic function of cells responsible for insulin action. Since fat tissue is considered as the main storing site of these pollutants in the body, many studies have been conducted on their effects on inflammatory pathways in adiposytes (Arsenescu et al. 2008; Baker et al. 2013; Ibrahim et al. 2011; Imbeault et al. 2012). But there is enough evidence that metabolic disorders, particularly IR, are accompanied with generally activated inflammatory cascades in different organs including fat, liver, muscle, and hypothalamus (Dali-Youcef et al. 2013; de Luca and Olefsky 2008; Odegaard and Chawla 2013; Park et al. 2014; Wieser et al. 2013). Since cellular metabolism in these organs is correlated by a hormonal network, dysfunction of one by an external intervention cannot be limited to itself and influencing the other partners in the network can result in metabolic diseases.

In this review, the evidences on the link between exposure to POPs and incidence of obesity, insulin resistance, and type2 diabetes will be provided with regard to their potential in induction of inflammation as the mechanism of toxicity for development of these diseases.

2 Methods

In order to show the magnitude of exposure to POPs, a search was done on the human breast milk concentration of POPs in different regions of the world during the last decade and after eliminating duplicates or irrelevant papers, 65 ones were reviewed and concluded in Table 1. For the association of POPs and insulin

Table 1 Breast milk concentration (ng/g milk fat) of some POPs in different region of the world

Region	DDTs	HCHs	HCB	CHLs	PCBs	PCDD/Fs	PBDEs	Time	Study
Australia	319	80	18	20	6.64[a]	0.095	11.1	2002–2004	Harden et al. (2007); Mueller et al. (2008); Toms et al. (2007)
							8.5	2007–2008	Toms et al. (2009)
Japan	340	110	14	80	19.13[a]	0.17	2.54	2001–2004	Kunisue et al. (2006); Eslami et al. (2006)
					13[a]	0.084		2002–2005	Todaka et al. (2010)
	170	140	13	31	110		1.5	2007–2008	Haraguchi et al. (2009)
		63			112			2007–2008	Fujii et al. (2012)
Korea	180	110	13	14	61		3.7	2007–2008	Haraguchi et al. (2009)
		50			63			2007–2008	Fujii et al. (2012)
China	870	550	56	6.7	28; 3700[a]	0.052		2002	Kunisue et al. (2004)
	2100	1400	81	16	42; 4000[a]	0.062		2002	Kunisue et al. (2004)
	3330	108	97			19[b]		2002	Sun et al. (2005)
	1916	21	84			13[b]	3.5	2003	Sun et al. (2005); Bi et al. (2006)
	1268	262	40		292		1.11	2003–2005	Zhao et al. (2007); Li et al. (2008)
	750	420	28		3.15[b]	11.5[b]		2006–2007	Leng et al. (2009)
	582	228	33	3	1.69[b]	3.73[b]		2007	Zhou et al. (2011); Li et al. (2009)
	1300	570	86	3.8	56		1.9	2007–2008	Haraguchi et al. (2009)
		688			46			2007–2008	Fujii et al. (2012)
Hong Kong	1500	942	21	6.1			3.4	2001–2002	Hui et al. (2008); Hedley et al. (2010)
	3099				49			2005	Tsang et al. (2011)
Taiwan	333					0.169	3.9	2000–2001	Chao et al. (2004, 2006, 2007); Wang et al. (2008)
						7.4[b]			Chao et al. (2005)

Vietnam	2300	58	3.9	6.9	79			2000–2001	Minh et al. (2004)
	1200	140	7	0.75	84		0.42	2007–2008	Haraguchi et al. (2009)
Indonesia	820	8.3	1.6	1.7	17		2.5	2001–2003	Sudaryanto et al. (2006, 2008)
	440	7.9	1.6	1.7	20		2.6	2001–2003	Sudaryanto et al. (2006, 2008)
	910	5.5	2.3	2.3	25		1.4	2001–2003	Sudaryanto et al. (2006, 2008)
India	1200	4500	4.2	7.3	34			2002–2003	Subramanian et al. (2007)
	1500	340	3.2	2.6	23			2004–2006	Devanathan et al. (2009)
	1100	670	4.4	2.8	40	0.62		2004–2006	Devanathan et al. (2009)
	175	127			102.7[a]			2005	Kumar et al. (2006); Someya et al. (2010)
	2870	2330						2010	Mishra and Sharma (2011)
	3210	2720						2010	Mishra and Sharma (2011)
Iran	2534	3970	972		1501			2006	Behrooz et al. (2009)
Turkey	32	2.2	0.3	0.39	1.08		0.008	2003	Erdogrul et al. (2004)
	338	37	5.44		8			2009	Cok et al. (2012)
Spain					255; 0.07[a]		2.4	2002	Schuhmacher et al. (2007)
	238				111	0.082	0.33	2004	Bordajandi et al. (2008)
					0.28	0.092	2.5	2007	Schuhmacher et al. (2009)
Italy	1024		38		532		4.1	2000–2001	Guerranti et al. (2011); Ingelido et al. (2007)
Poland	1195	20	22		114			2000–2001	Szyrwińska and Lulek (2007)
	868	14	32		153		2.5	2004	Jaraczewska et al. (2006)
Czech							1.73	2003	Kazda et al. (2004)
Germany	180				230; 21.2[a]	0.08	1.96	2005	Raab et al. (2008)
	125	21	26		202			2006	Zietz et al. (2008)
Latvia	219	54	25		141; 128[a]	0.089		2006	Bake et al. (2007)
Belgium	156	12	15		89	10.31[b]	2.01	2006	Colles et al. (2008)
	196	8.9	9.6		76; 5.9[a]	8.4[b]	2.03	2009–2010	Croes et al. (2012)

(continued)

Table 1 (continued)

Region	DDTs	HCHs	HCB	CHLs	PCBs	PCDD/Fs	PBDEs	Time	Study
Norway	110	13	18	14.3	170		4.1	2000–2001	Polder et al. (2008b)
	53	5.4	11	3	112			2002–2006	Polder et al. (2009)
			12					2003–2006	Eggesbo et al. (2009)
Sweden					136[a]	0.018	3.5	1996–2006	Lignell et al. (2009)
Russia	900	235	65	22				2000–2002	Polder et al. (2008a)
	1037	186	58	21				2000–2002	Polder et al. (2008a)
	1098	163	46	22				2000–2002	Polder et al. (2008a)
Ghana	78	46	4.9					2005	Ntow et al. (2008)
Ethiopia	12,683							2010	Gebremichael et al. (2013)
Tunisia	1931	65	85		196			2003–2005	Ennaceur et al. (2008)
	1163		286		331			2010	Hassine et al. (2012)
South Africa	4797							2002	Bouwman et al. (2006)
	6320	12	1.9		10		1.7	2004	Darnerud et al. (2011)
	9500							2008	Bouwman et al. (2012)
	11,000							2008	Bouwman et al. (2012)
	18,000							2008	Bouwman et al. (2012)
Brazil	492							2001–2002	Azeredo et al. (2008)
US and Canada					147		95.6	2003	She et al. (2007)

[a]Dioxin-like PCBs
[b]Unite as WHO TEQ pg/g

resistance, Google Scholar, PubMed, and Scopus were searched between the years 1960–2012 for the keywords "persistent organic pollutant, organochlorine pesticide, polychlorinated biphenyl, obesity, insulin resistance, diabetes, metabolic syndrome". Around 200 articles were found in primary search but after elimination of duplicates or irrelevant papers, only 39 records remained to be reviewed (Tables 2, 3, and 4).

3 Results

3.1 POPs and Metabolic Disorders of Insulin Resistance

3.1.1 Epidemiological Evidences

The U.S National Health and Nutrition Examination Survey 1999–2002 conducted by the Centers for Disease Control and Prevention was among the first comprehensive programs whose results cleared the association between serum concentration of POPs and metabolic disorders including insulin resistance, diabetes, and metabolic syndrome. Analyzing those data, Lee et al. reported that there is a significant relation between diabetes and serum concentration of six selected POPs including Oxychlordane, 2,2,4,4,5,5-hexachlorobiphenyl (PCB153), trans-Nonachlor, 1,2,3,4,6,7,8-heptachlorodibenzo-p-dioxin (HpCDD), 1,2,3,4,6,7,8,9-octachlorodibenzo-p-dioxin (OCDD), and *p,p′*-dichlorodiphenyltrichloroethane (DDE) (Lee et al. 2006a, b). Further, they reported an association between exposure to some POPs, particularly organochlorine pesticides, and different components of metabolic syndrome in non-diabetic adults (Lee et al. 2007b). The prevalence of insulin resistance in non-diabetic adults was also found to be associated with serum concentration of organochlorine pesticides and nondioxin-like polychlorinated biphenyls (PCBs) suggesting the involvement of POPs in developing diabetes by increasing its main risk factor obesity (Lee et al. 2007a).

A cross-sectional analysis of the results from the National Health and Nutrition Examination Survey 2003–2004 examining 1367 adults, indicated that brominated flame retardants (BFRs), such as polybrominated diphenyl ethers (PBDEs) or polybrominated biphenyls (PBBs) with respect to their storing in adipose tissue may be involved in development of diabetes and metabolic syndrome (Lim et al. 2008).

Lee et al. (2010) conducted a prospective nested case control study on the Coronary Artery Risk Development in Young Adults (CARDIA) cohort to predict whether exposure to low dose of POPs can predict the incidence of type2 diabetes. Their results showed that there is a nonlinear association between diabetes and low dose of POPs similar to the current exposure, and the highest risk was reported for mirex, oxychlordane, trans-nonachlor, highly chlorinated PCBs, and PBB153-a (Lee et al. 2010). Following this study the authors did another nested case control analysis on the control group without diabetes to examine the development of dysmetabolic features and found that low level exposure to POPs in general population can cause obesity, dyslipidemia, and insulin resistance (Lee et al. 2011).

Table 2 Epidemiological evidences on the association of exposure to POPs with metabolic disorders

Study	Type	N	Sample	POPs	Pathology
Lee et al. (2006a, b)	Cross-sectional	2016	Serum	OCPs, PCBs	Diabetes
Lee et al. (2007a)	Cross-sectional	749	Serum	OCPs, PCBs	IR
Lee et al. (2007b)	Cross-sectional	721	Serum	OCPs, PCBs	Metabolic syndrome
Lim et al. (2008)	Cross-sectional	1367	Serum	PBDEs	Diabetes, Metabolic syndrome
Lee et al. (2010)	Nested case control	180	Serum	OCPs, PCBs	Type2 diabetes
Lee et al. (2011)	Nested case control	90	Serum	OCPs, PCBs	Obesity, IR, dyslipidemia
Lee et al. (2012)	Prospective cross sectional	970	Plasma	PCBs, p,p′-DDE, dioxin	Obesity
Uemura et al. (2009)	Cross-sectional	1347	Blood	Dioxin, dioxin-like PCBs	Metabolic syndrome
Park et al. (2010)	Case control	100	Serum	β-HCH, heptachlor epoxide	Metabolic syndrome
Persky et al. (2011)	Cohort-retrospective	118	Serum	PCBs	Diabetes
Persky et al. (2012)	Cohort-retrospective	63	Serum	PCBs	Diabetes
Dirinck et al. (2011)	Cross sectional	145	Serum	β-HCH	Obesity
Dirinck et al. (2014)	Case control	195	Serum	PCBs, DDTs	Obesity, glucose intolerance
Faerch et al. (2012)	Cross sectional	148	Serum		Diabetes and prediabetes
Kim et al. (2012a, b)	Cross sectional	748	Serum	OCPs, PCBs	IR
Arrebola et al. (2013)	Case control	386	Serum, fat	OCPs, PCBs	Type 2 diabetes
Pal et al. (2013)	Case control	72	Plasma	Common POPs	Diabetes
Rignell-Hydbom et al. (2007)	Cross sectional	503	Serum	DDTs, dl-PCBs	Diabetes
Turyk et al. (2009)	Cross sectional	544	Serum	DDTs, PCBs	Diabetes

N number of samples

Table 3 Experimental evidences on the association of exposure to POPs with metabolic disorders

Study	Model	Sample type	POP	Results
Wang et al. (2010b)	In vitro	human umbilical vascular endothelial cells	PCB77	↓ insulin signaling
Ruzzin et al. (2010)	In vitro and In vivo	Rat and differentiated adipocytes	Farmed salmon containing POPs	Obesity, hepatosteatosis, IR, ↓Insig-1 and Lpin-1
Ibrahim et al. (2011)	In vivo	Mice, adipose tissue	Farmed salmon containing POPs	Obesity, IR
Baker et al. (2013)		Mice, adipose tissue	PCBs	IR
Gray et al. (2013)	In vivo	Mice	Aroclor 1254	Obesity, IR, hyperinsulinemia
Wahlang et al. (2013)	In vivo	Mice	PCB153	Visceral obesity, hepatosteatosis,
Arsenescu et al. (2008)	In vitro and In vivo	Mice, adipocytes	PCB77	Obesity, ↑ adipocyte differentiation
Fang et al. (2012)	In vitro and In vivo	Rat, hepatocytes	perfluorononanoic acid (PFNA)	hepatic lipid accumulation
Howell et al. (2011)	In vitro	NIH3T3-L1 cell	DDE	↑ leptin, resistin, and adiponectin

A cross-sectional and prospective analysis on participants of the Prospective Investigation of the Vasculature in Uppsala Seniors (PIVUS) indicated that low level exposure to POPs, particularly less chlorinated PCBs, p,p′-DDE, and dioxin, were associated with abdominal obesity in elderly (Lee et al. 2012). Recently, the POPs in association with markers of glucose homeostasis was studied in two First Nations communities in northern Ontario where the most people use wild food and the results confirmed the higher plasma concentration of POPs in diabetic than non-diabetic individuals (Ritter et al. 2007).

The result of a cross sectional study on the general population not occupationally exposed to POPs in Japan showed that there is a relation between prevalence of metabolic syndrome and body burden levels of dioxins and dioxin-like PCBs (Uemura et al. 2009). The serum concentrations of organochlorine pesticides, beta-hexachlorocyclohexane and heptachlor epoxide, were evaluated in a community based health survey and reported to be associated with the components of metabolic syndrome (Park et al. 2010).

Persky et al. (2011, 2012) conducted two studies on men and women previously employed at a capacitor manufacturing plant and reported that there is a link between serum concentration of PCBs and diabetes (Persky et al. 2011, 2012).

A positive relationship between high serum level of the organochlorine pesticide, beta-hexachlorocyclohexane, and BMI and HOMA (IR) was established by Dirinck et al. cross-sectionally investigating the association of obesity with serum

Table 4 Investigations evidencing proinflammatory properties of POPs

Study	Model	Sample type	POP	Results
Hennig et al. (2002)	In vitro	Porcine endothelial cells	Coplanar PCBs	Oxidative stress, ↑ NF-κB and IL-6
Kwon et al. (2002)	In vitro	Human leukemic mast cells	PCB153	↑ COX-2, IL-6, NF-κB
Choi et al. (2003)	In vitro	Human vascular endothelial cells	PCB104	Oxidative stress, Inflammation
Nyska et al. (2004)	In vivo	Rat, pancreas	Dioxin, dioxin-like compounds	Chronic active inflammation
Arsenescu et al. (2008)	In vitro and In vivo	Mice, adipocytes	PCB77	↑Inflammatory adipokines
Fang et al. (2012)	In vitro and In vivo	Rat, hepatocytes	Perfluorononanoic acid (PFNA)	↑ Inflammatory cytokines
Kim et al. (2012a, b)	In vitro	Human multipotent adipose-derived stem cells	TCDD, PCB126, PCB153	Modulation of inflammatory genes
Imbeault et al. (2012)	Epidemiologic	Canadian First Nations communities	Mirex, Aroclor 1260, PCBs (153, 170, 180, 187)	↑ Plasma cytokines
Wang et al. (2010b)	In vitro	Human umbilical vascular endothelial cells	PCB77	↑ IL-6, TNF-α, NF-κB
Ibrahim et al. (2011)	In vivo	Mice, adipose tissue	Farmed salmon containing POPs	↑TNF-α, macrophage infiltration
Baker et al. (2013)	In vivo	Mice, adipose tissue	PCBs	↑ TNF-α

level of POPs (Dirinck et al. 2011). Another cross-sectional study carried out on Danish middle-aged people showed that prediabetic and diabetic individuals have higher serum concentrations of some POPs with regard to their potential to change substrate oxidation patterns (Faerch et al. 2012).

Kim et al. (2012a, b) conducted a cross-sectional study to evaluate the relation between serum concentration of POPs, and the markers of inflammation and, C-reactive protein (CRP) and (HOMA-IR), respectively in non-diabetic adults. The correlation between POPs and CRP was significant just in case of organochlorine pesticides, and a significant association was found between CRP and insulin resistance when the concentration of organochlorine pesticides or PCBs was high (Kim et al. 2012a).

3.1.2 Experimental Evidences

There are several experimental studies with mechanistic insights into the metabolic disrupting effects of POPs and regarding insulin resistance most of them directed inflammatory pathways. Wang et al. (2010b) examined the effect of PCB77 on insulin signaling and inflammatory response in human umbilical vascular endothelial cells (HUVEC) and reported blockade of insulin-activated Akt signaling pathway. Induction of IL-6, TNF-α along with increased activity of the transcription factor NFkappaB were the other findings of the study in which pretreatment of cells with an antibody counteracting TNF-α improved the effect of insulin on Akt signaling cascade (Wang et al. 2010b). The results of an experimental study carried out on rats exposed to common POPs found in the food chain implied on development of abdominal obesity, hepatosteatosis and insulin resistance. Following this observation the *in vitro* part of the study conducted on differentiated adipocytes, also confirmed inhibition of insulin signaling accompanied with down-regulation of two main regulators of lipid metabolism, insulin-induced gene-1 (Insig-1) and Lpin1, by POPs particularly organochlorine pesticides (Ruzzin et al. 2010). Another study showed that mice chronically fed with famed salmon fillet had higher body burden of POPs along with visceral obesity, glucose intolerance, and insulin resistance with regard to reduced ability of insulin in stimulating Akt signaling. Those animals had also a higher expression level of the pro-inflammatory cytokine TNF-α and macrophage infiltration in adipose tissues (Ibrahim et al. 2011). Baker et al. (2013) examined the effect of PCBs on glucose homeostasis in mice and indicated that these chemicals impair glucose and insulin tolerance through the aryl hydrocarbon receptor (AhR). They also reported that the level of PCBs and expression level of TNF-α were more in adipose tissues than any other organ involved in glucose homeostasis (Baker et al. 2013). Chronic exposure of mice to a PCB, Aroclor 1254, has been reported to exacerbate obesity related insulin resistance and cause hyperinsulinemia in both lean and obese mice (Gray et al. 2013). PCB 153, a known obesogen, has also been shown to cause visceral obesity, hepatosteatosis, and altered profile of adipocytokines' expression in mice (Wahlang et al. 2013).

3.2 POPs and Inflammation

Many studies have been done in different models regarding POPs' effects on the components of the immune system focusing on inflammation by which the link between exposure to these chemicals and metabolic disorders can be uncovered. Coplanar PCBs were examined in an experiment conducted on porcine endothelial cells and the results implicated on induction of oxidative stress, and increased activity of NF-κB and production of IL-6. Proinflammatory properties of these chemicals were thought to be mediated through AhR for which coplanar PCBs act as agonists (Hennig et al. 2002). In another study on human leukemic mast cells, PCB153, which is not an agonist of AhR, was shown to increase the mRNA expression of cyclooxygenase-2 (COX-2) and IL-6 through activating NF-κB pathway (Kwon et al. 2002). PCB104 as a non-coplanar PCB that does not activate the AhR receptor was also reported to induce oxidative stress and up-regulate proinflammatory components in human vascular endothelial cells (Choi et al. 2003). Nyska et al. (2004) examined the pancreatic toxicity of POPs in rats administered dioxin and dioxin-like compounds for 2 years and indicated the higher incidence of chronic active inflammation in the exocrine pancreas of treated animals (Nyska et al. 2004). The effect of PCB77 on adipose tissue was examined both *in vitro* and *in vivo* by Arsenescu et al. (2008) and increased differentiation of adipocytes and expression of proinflammatory adipokines were concluded. They also reported that PCB77 induced obesity which was abolished when an antagonist of AhR was administered (Arsenescu et al. 2008). A study conducted on rat hepatocytes and Kupffer cells both *in vitro* and *in vivo* revealed that perfluoro-nonanoic acid (PFNA), a persistent organic pollutant, increased the expression of inflammatory cytokines and changed the profile of lipid metabolism leading to lipid accumulation in the liver (Fang et al. 2012). Kim et al. (2012a, b) conducted a large-scale gene expression analysis to determine the major targets of 2,3,7,8-tetrachlor-odibenzo-p-dioxin (TCDD), PCB126, and PCB153 in adipose tissue. The results showed that POPs acting as AhR ligand target the adipose tissue mainly by modulating those genes involved in inflammatory pathway (Kim et al. 2012b). Furthermore, in an epidemiologic study conducted on Canadian First Nations communities, a significant association was detected between plasma concentration of POPs and increased level of cytokines (Imbeault et al. 2012).

4 Discussion

Investigations regarding the relation between exposure to POPs and IR based disorders are limited to the last decade after the Stockholm Convention on Persistent Organic Pollutants in 2001 which alarmed the environmental and health hazards of these chemicals. Since then the list of POPs has been expanded and some other chemicals were joined creating a need for updated research on their metabolic disrupting effects.

Table 1 represents the breast milk levels of some POPs in different regions of the world during the last decade. Although, POPs' concentration in human milk shows an overall decrease compared with their peak in 1970–1990s, there are still some regions in which significant levels of POPs in breast milk can be found. Measurement of POPs in breast milk is highly considerable for two reason; first, lipophilic chemicals tend to be secreted into the milk so measurement of milk concentration can be a good indicator for body burden of POPs and second, the morphology and function of organ systems in the body are so susceptible to chemical exposure during gestational period.

Since POPs were used to be widely produced and utilized as pesticides and industrial intermediates, retrospective cohort studies can be helpful in finding the link between this kind of exposures and metabolic diseases. As presented in Table 2, there are two of these types of studies conducted by Persky et al. (2011, 2012) reporting the higher incidence of diabetes in workers who were occupationally exposed to PCBs. The other epidemiologic evidences are case-control or cross-sectional reporting the higher levels of different POPs in biological samples, mostly blood, taken from individuals who are obese or suffer from diabetes, or metabolic syndrome. The common feature of all these complications is IR which itself has been directly detected in some above mentioned studies on POPs. However, most epidemiologic studies have confirmed the existence of a correlation between body burden of POPs and higher prevalence of IR related metabolic diseases. Furthermore, there are experimental studies both *in vitro* and *in vivo* evaluating the metabolic disrupting effects of different POPs and most of them recorded IR along with obesity and the other features of unbalanced homeostasis of glucose and lipid (Table 3). When the responsiveness of biologic systems to insulin decreases, hyperglycemia may appear or not, but altered lipid metabolism will commonly and primarily become manifested by elevated synthesis and accumulation of lipid in adipose tissues (as obesity), and in the liver (as fatty liver disease). Table 3 shows that POPs experimentally can increase lipid accumulation and adipocyte differentiation in adipose tissues leading to visceral obesity. In addition, hepatic steatosis due to lipid accumulation in hepatocytes has been experimentally reported to be promoted by POPs (Table 3). Some experimental studies have had mechanistic insights into the link between exposure to POPs and IR related metabolic disruption. One of the main mechanisms which have been confirmed to promote or accompany the IR is inflammation and related pathways in insulin target organs including liver, adipose tissues and muscles. Different POPs have been tested and these studies indicated that increase the level of inflammatory cytokines like TNF-α and IL-6, and activate the transcription factor involved in inflammation, NF-κB, in hepatocytes, adipocytes, and endothelial cells (Table 4).

In fact, these mechanistic evidences on pro-inflammatory properties of POPs create the assumption that obesity can be not only a risk factor for IR, but also the outcome of IR. Stimulated inflammatory response works as an adaptive mechanism against POPs stored in adipose tissues by attenuating responsiveness of adipocytes to glucose but not lipid storing action of insulin, so that the rate of lipogenesis in adipocytes increases. In the long term such a vicious cycle can result in both obesity and IR which usually coexist with each other.

On the other side, obese people have enlarged adipocytes allowing further accumulation and storage of lipophilic POPs so that adipose tissues can become an internal source for releasing these chemicals chronically. Accumulation of POPs in adipose tissues can also disturb the normal homeostasis of adipocytes leading to altered expression and function of inflammatory cytokines and adipokines. Inflammatory response is usually accompanied with attenuation of insulin signaling in order to adapt the changes. Similarly, such a resistance to insulin action can be induced by inflammation in the other insulin responding organs, liver and muscles (Fig. 1).

Although the controversy over the casual link between IR and obesity is repeated on the association of POPs with obesity raising the old question, which one arises first and causes the other, accompanying POPs with IR related metabolic disorder cannot be disclaimed according to the evidences. The common metabolic disease

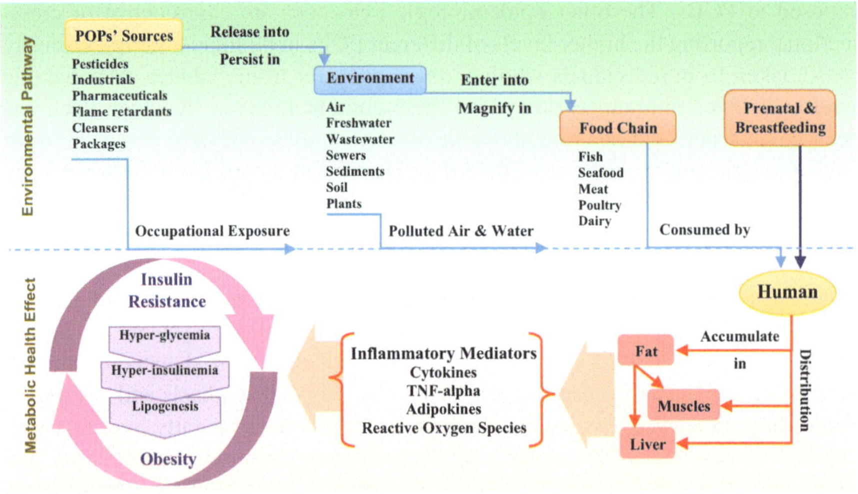

Fig. 1 A schematic diagram representing environmental and human exposure to POPs leading to insulin resistance related metabolic disorders

developed by IR is diabetes from which 347 million people suffer worldwide, and its burden increasingly grows so that is projected to be the 7th leading cause of death in 2030 (www.who.net). Considering the overall afflictions indirectly caused by the other IR-related metabolic disorders like dyslipidemia, metabolic syndrome, and cardiovascular diseases, mentioned statistics may increase by several ten times.

5 Conclusion

The proficiency of the insulin signaling cascade in enabling biological systems to cope with changes has made it a very sensitive target for environmental exposures, in particular for POPs. The concern over POPs contamination is not only due to their emissions, but also their persistence in the environment. This cause them to bio-accumulate in the food chain and storage in the human body which chronically can lead to perturb the health of related target systems particularly endocrine and immune components. In this way, POPs induced inflammation and production of oxidative and inflammatory mediators in insulin target tissues including fat, liver, and muscles can result in attenuation of cellular insulin signaling, which can be manifested as insulin resistance. Although, the Stockholm Convention on Persistent Organic Pollutants in 2001 limited the production and use of POPs somewhat, the problem of their elimination from the environment remained to be solved. This should be noteworthy for agencies responsible for environmental protection in order to plan for eliminating in addition to limiting the emission of these chemicals.

Acknowledgment Thank to the Research Council of Ardabil University of Medical Sciences.

Conflict of Interests
There is no conflict of interest.

References

Arrebola JP et al (2013) Adipose tissue concentrations of persistent organic pollutants and prevalence of type 2 diabetes in adults from Southern Spain. Environ Res 122:31–37. doi:10.1016/j.envres.2012.12.001

Arsenescu V, Arsenescu RI, King V, Swanson H, Cassis LA (2008) Polychlorinated biphenyl-77 induces adipocyte differentiation and proinflammatory adipokines and promotes obesity and atherosclerosis. Environ Health Perspect 116:761–768. doi:10.1289/ehp.10554

Azeredo A et al (2008) DDT and its metabolites in breast milk from the Madeira River basin in the Amazon, Brazil. Chemosphere 73:S246–S251. doi:10.1016/j.chemosphere.2007.04.090

Bake MA, Linnika Z, Sudmalis P, Kocan A, Jursa S, Pike A, Ruce M (2007) Assessment of the exposure of breast milk to persistent organic pollutants in Latvia. Int J Hyg Environ Health 210:483–489. doi:10.1016/j.ijheh.2007.01.016

Baker NA et al (2013) Coplanar polychlorinated biphenyls impair glucose homeostasis in lean C57BL/6 mice and mitigate beneficial effects of weight loss on glucose homeostasis in obese mice. Environ Health Perspect 121:105–110. doi:10.1289/ehp.1205421

Behrooz RD, Sari AE, Bahramifar N, Ghasempouri SM (2009) Organochlorine pesticide and polychlorinated biphenyl residues in human milk from the Southern Coast of Caspian Sea, Iran. Chemosphere 74:931–937. doi:10.1016/j.chemosphere.2008.10.014

Bi X et al (2006) Polybrominated diphenyl ethers in South China maternal and fetal blood and breast milk. Environ Pollut 144:1024–1030. doi:10.1016/j.envpol.2005.12.056

Bordajandi LR, Abad E, González MJ (2008) Occurrence of PCBs, PCDD/Fs, PBDEs and DDTs in Spanish breast milk: enantiomeric fraction of chiral PCBs. Chemosphere 70:567–575. doi:10.1016/j.chemosphere.2007.07.019

Bouwman H, Sereda B, Meinhardt HM (2006) Simultaneous presence of DDT and pyrethroid residues in human breast milk from a malaria endemic area in South Africa. Environ Pollut 144:902–917. doi:10.1016/j.envpol.2006.02.002

Bouwman H, Kylin H, Sereda B, Bornman R (2012) High levels of DDT in breast milk: intake, risk, lactation duration, and involvement of gender. Environ Pollut 170:63–70. doi:10.1016/j.envpol.2012.06.009

Chao HR, Wang SL, Lee CC, Yu HY, Lu YK, Papke O (2004) Level of polychlorinated dibenzo-p-dioxins, dibenzofurans and biphenyls (PCDD/Fs, PCBs) in human milk and the input to infant body burden. Food Chem Toxicol 42:1299–1308. doi:10.1016/j.fct.2004.03.008

Chao HR, Wang SL, Su PH, Yu HY, Yu ST, Papke O (2005) Levels of polychlorinated dibenzo-p-dioxins and dibenzofurans in primipara breast milk from Taiwan: estimation of dioxins and furans intake for breastfed infants. J Hazard Mater 121:1–10. doi:10.1016/j.jhazmat.2004.09.021

Chao HR, Wang SL, Lin TC, Chung XH (2006) Levels of organochlorine pesticides in human milk from central Taiwan. Chemosphere 62:1774–1785. doi:10.1016/j.chemosphere.2005.07.036

Chao HR, Wang SL, Lee WJ, Wang YF, Päpke O (2007) Levels of polybrominated diphenyl ethers (PBDEs) in breast milk from central Taiwan and their relation to infant birth outcome and maternal menstruation effects. Environ Int 33:239–245. doi:10.1016/j.envint.2006.09.013

Choi W, Eum SY, Lee YW, Hennig B, Robertson LW, Toborek M (2003) PCB 104-induced proinflammatory reactions in human vascular endothelial cells: relationship to cancer metastasis and atherogenesis. Toxicol Sci 75:47–56. doi:10.1093/toxsci/kfg149

Cok I, Mazmanci B, Mazmanci MA, Turgut C, Henkelmann B, Schramm KW (2012) Analysis of human milk to assess exposure to PAHs, PCBs and organochlorine pesticides in the vicinity Mediterranean city Mersin, Turkey. Environ Int 40:63–69. doi:10.1016/j.envint.2011.11.012

Colles A et al (2008) Fourth WHO-coordinated survey of human milk for persistent organic pollutants (POPs): Belgian results. Chemosphere 73:907–914. doi:10.1016/j.chemosphere.2008.07.002

Croes K et al (2012) Persistent organic pollutants (POPs) in human milk: a biomonitoring study in rural areas of Flanders (Belgium). Chemosphere 89:988–994. doi:10.1016/j.chemosphere.2012.06.058

Dali-Youcef N, Mecili M, Ricci R, Andres E (2013) Metabolic inflammation: connecting obesity and insulin resistance. Ann Med 45:242–253. doi:10.3109/07853890.2012.705015

Darnerud PO et al (2011) Levels of brominated flame retardants and other persistent organic pollutants in breast milk samples from Limpopo Province, South Africa. Sci Total Environ 409:4048–4053. doi:10.1016/j.scitotenv.2011.05.054

de Luca C, Olefsky JM (2008) Inflammation and insulin resistance. FEBS Lett 582:97–105. doi:10.1016/j.febslet.2007.11.057

Devanathan G et al (2009) Persistent organochlorines in human breast milk from major metropolitan cities in India. Environ Pollut 157:148–154. doi:10.1016/j.envpol.2008.07.011

Dirinck E et al (2011) Obesity and persistent organic pollutants: possible obesogenic effect of organochlorine pesticides and polychlorinated biphenyls. Obesity (Silver Spring) 19:709–714. doi:10.1038/oby.2010.133

Dirinck EL, Dirtu AC, Govindan M, Covaci A, Van Gaal LF, Jorens PG (2014) Exposure to persistent organic pollutants: relationship with abnormal glucose metabolism and visceral adiposity. Diabetes Care 37:1951–1958. doi:10.2337/dc13-2329

Eggesbo M et al (2009) Levels of hexachlorobenzene (HCB) in breast milk in relation to birth weight in a Norwegian cohort. Environ Res 109:559–566. doi:10.1016/j.envres.2009.04.001

Ennaceur S, Gandoura N, Driss MR (2008) Distribution of polychlorinated biphenyls and organochlorine pesticides in human breast milk from various locations in Tunisia: levels of contamination, influencing factors, and infant risk assessment. Environ Res 108:86–93. doi:10.1016/j.envres.2008.05.005

Erdogrul O, Covaci A, Kurtul N, Schepens P (2004) Levels of organohalogenated persistent pollutants in human milk from Kahramanmaras region, Turkey. Environ Int 30:659–666. doi:10.1016/j.envint.2003.12.004

Eslami B et al (2006) Large-scale evaluation of the current level of polybrominated diphenyl ethers (PBDEs) in breast milk from 13 regions of Japan. Chemosphere 63:554–561. doi:10.1016/j.chemosphere.2005.09.067

Faerch K, Hojlund K, Vind BF, Vaag A, Dalgard C, Nielsen F, Grandjean P (2012) Increased serum concentrations of persistent organic pollutants among prediabetic individuals: potential role of altered substrate oxidation patterns. J Clin Endocrinol Metab 97:E1705–E1713. doi:10.1210/jc.2012-1342

Fang X, Gao G, Xue H, Zhang X, Wang H (2012) In vitro and in vivo studies of the toxic effects of perfluorononanoic acid on rat hepatocytes and Kupffer cells. Environ Toxicol Pharmacol 34:484–494. doi:10.1016/j.etap.2012.06.011

Fujii Y, Ito Y, Harada KH, Hitomi T, Koizumi A, Haraguchi K (2012) Comparative survey of levels of chlorinated cyclodiene pesticides in breast milk from some cities of China, Korea and Japan. Chemosphere 89:452–457. doi:10.1016/j.chemosphere.2012.05.098

Gebremichael S, Birhanu T, Tessema DA (2013) Analysis of organochlorine pesticide residues in human and cow's milk in the towns of Asendabo, Serbo and Jimma in South-Western Ethiopia. Chemosphere 90:1652–1657. doi:10.1016/j.chemosphere.2012.09.008

Gray SL, Shaw AC, Gagne AX, Chan HM (2013) Chronic exposure to PCBs (Aroclor 1254) exacerbates obesity-induced insulin resistance and hyperinsulinemia in mice. J Toxicol Environ Health A 76:701–715. doi:10.1080/15287394.2013.796503

Guerranti C, Palmieri M, Mariottini M, Focardi SE (2011) Persistent organic pollutants in human milk from central Italy: levels and time trends. ISRN Toxicol 2011:107514. doi:10.5402/2011/107514

Halldorsson T (2012) Dietary exposures to persistent organic pollutants and fetal growth. In: Preedy VR (ed) Handbook of growth and growth monitoring in health and disease. Springer, New York, NY, pp 2559–2578. doi:10.1007/978-1-4419-1795-9_152

Haraguchi K et al (2009) Levels and regional trends of persistent organochlorines and polybrominated diphenyl ethers in Asian breast milk demonstrate POPs signatures unique to individual countries. Environ Int 35:1072–1079. doi:10.1016/j.envint.2009.06.003

Harden FA, Toms LML, Symons R, Fürst P, Berry Y, Müller JF (2007) Evaluation of dioxin-like chemicals in pooled human milk samples collected in Australia. Chemosphere 67:S325–S333. doi:10.1016/j.chemosphere.2006.05.126

Hassine SB, Ameur WB, Gandoura N, Driss MR (2012) Determination of chlorinated pesticides, polychlorinated biphenyls, and polybrominated diphenyl ethers in human milk from Bizerte (Tunisia) in 2010. Chemosphere 89:369–377. doi:10.1016/j.chemosphere.2012.05.035

Hedley AJ et al (2010) Residues of persistent organic pollutants (POPs) in human milk in Hong Kong. Chemosphere 79:259–265. doi:10.1016/j.chemosphere.2010.01.047

Hennig B, Meerarani P, Slim R, Toborek M, Daugherty A, Silverstone AE, Robertson LW (2002) Proinflammatory properties of coplanar PCBs: in vitro and in vivo evidence. Toxicol Appl Pharmacol 181:174–183, S0041008X02994081 [pii]

Howell G 3rd, Mangum L (2011) Exposure to bioaccumulative organochlorine compounds alters adipogenesis, fatty acid uptake, and adipokine production in NIH3T3-L1 cells. Toxicol In Vitro 25:394–402. doi: 10.1016/j.tiv.2010.10.015

Hui LL et al (2008) DDT levels in human milk in Hong Kong, 2001-02. Chemosphere 73:50–55. doi:10.1016/j.chemosphere.2008.05.045

Ibrahim MM et al (2011) Chronic consumption of farmed salmon containing persistent organic pollutants causes insulin resistance and obesity in mice. PLoS One 6:e25170. doi:10.1371/journal.pone.0025170

Imbeault P et al (2012) Dysregulation of cytokine response in Canadian First Nations communities: is there an association with persistent organic pollutant levels? PLoS One 7:e39931. doi:10.1371/journal.pone.0039931

Ingelido AM et al (2007) Polychlorinated biphenyls (PCBs) and polybrominated diphenyl ethers (PBDEs) in milk from Italian women living in Rome and Venice. Chemosphere 67:S301–S306. doi:10.1016/j.chemosphere.2006.05.111

Jaraczewska K, Lulek J, Covaci A, Voorspoels S, Kaluba-Skotarczak A, Drews K, Schepens P (2006) Distribution of polychlorinated biphenyls, organochlorine pesticides and polybrominated diphenyl ethers in human umbilical cord serum, maternal serum and milk from Wielkopolska region, Poland. Sci Total Environ 372:20–31. doi:10.1016/j.scitotenv.2006.03.030

Jurjanz S, Rychen G, Feidt C (2008) Dairy livestock exposure to persistent organic pollutants and their transfer to milk: a review. In: Faye B, Sinyavskiy Y (eds) Impact of pollution on animal products, NATO science for peace and security series. Series C: environmental security. Springer, Netherlands, pp 63–83. doi:10.1007/978-1-4020-8359-4_7

Kazda R, Hajšlová J, Poustka J, Čajka T (2004) Determination of polybrominated diphenyl ethers in human milk samples in the Czech Republic: comparative study of negative chemical ionisation mass spectrometry and time-of-flight high-resolution mass spectrometry. Anal Chim Acta 520:237–243. doi:10.1016/j.aca.2004.04.069

Kim KS, Hong NS, Jacobs DR Jr, Lee DH (2012a) Interaction between persistent organic pollutants and c-reactive protein in estimating insulin resistance among non-diabetic adults. J Prev Med Public Health 45:62–69. doi:10.3961/jpmph.2012.45.2.62

Kim MJ et al (2012b) Inflammatory pathway genes belong to major targets of persistent organic pollutants in adipose cells. Environ Health Perspect 120:508–514. doi:10.1289/ehp.1104282

Kumar A, Dayal P, Shukla G, Singh G, Joseph PE (2006) DDT and HCH residue load in mother's breast milk: a survey of lactating mother's from remote villages in Agra region. Environ Int 32:248–251. doi:10.1016/j.envint.2005.08.019

Kunisue T, Someya M, Kayama F, Jin Y, Tanabe S (2004) Persistent organochlorines in human breast milk collected from primiparae in Dalian and Shenyang, China. Environ Pollut 131:381–392. doi:10.1016/j.envpol.2004.03.008

Kunisue T et al (2006) Contamination status of persistent organochlorines in human breast milk from Japan: recent levels and temporal trend. Chemosphere 64:1601–1608. doi:10.1016/j.chemosphere.2005.11.010

Kwon O et al (2002) Expression of cyclooxygenase-2 and pro-inflammatory cytokines induced by 2,2′,4,4′,5,5′-hexachlorobiphenyl (PCB 153) in human mast cells requires NF-kappa B activation. Biol Pharm Bull 25:1165–1168. doi:10.1248/bpb.25.1165

Lee DH (2012) Persistent organic pollutants and obesity-related metabolic dysfunction: focusing on type 2 diabetes. Epidemiol Health 34:e2012002. doi:10.4178/epih/e2012002

Lee DH, Jacobs DR, Porta M (2006a) Could low-level background exposure to persistent organic pollutants contribute to the social burden of type 2 diabetes? J Epidemiol Community Health 60:1006–1008. doi:10.1136/jech.2006.053389

Lee DH, Lee IK, Song K, Steffes M, Toscano W, Baker BA, Jacobs DR Jr (2006b) A strong dose-response relation between serum concentrations of persistent organic pollutants and diabetes: results from the National Health and Examination Survey 1999-2002. Diabetes Care 29:1638–1644. doi:10.2337/dc06-0543

Lee DH, Lee IK, Jin SH, Steffes M, Jacobs DR Jr (2007a) Association between serum concentrations of persistent organic pollutants and insulin resistance among nondiabetic adults: results from the National Health and Nutrition Examination Survey 1999-2002. Diabetes Care 30:622–628. doi:10.2337/dc06-2190

Lee DH, Lee IK, Porta M, Steffes M, Jacobs DR Jr (2007b) Relationship between serum concentrations of persistent organic pollutants and the prevalence of metabolic syndrome among non-diabetic adults: results from the National Health and Nutrition Examination Survey 1999-2002. Diabetologia 50:1841–1851. doi:10.1007/s00125-007-0755-4

Lee DH, Steffes MW, Sjodin A, Jones RS, Needham LL, Jacobs DR Jr (2010) Low dose of some persistent organic pollutants predicts type 2 diabetes: a nested case-control study. Environ Health Perspect 118:1235–1242. doi:10.1289/ehp.0901480

Lee DH, Steffes MW, Sjodin A, Jones RS, Needham LL, Jacobs DR Jr (2011) Low dose organochlorine pesticides and polychlorinated biphenyls predict obesity, dyslipidemia, and insulin resistance among people free of diabetes. PLoS One 6:e15977. doi:10.1371/journal.pone.0015977

Lee DH, Lind L, Jacobs DR Jr, Salihovic S, van Bavel B, Lind PM (2012) Associations of persistent organic pollutants with abdominal obesity in the elderly: the Prospective Investigation of the Vasculature in Uppsala Seniors (PIVUS) study. Environ Int 40:170–178. doi:10.1016/j.envint.2011.07.010

Leng J-H, Kayama F, Wang P-Y, Nakamura M, Nakata T, Wang Y (2009) Levels of persistent organic pollutants in human milk in two Chinese coastal cities, Tianjin and Yantai: influence of fish consumption. Chemosphere 75:634–639. doi:10.1016/j.chemosphere.2009.01.008

Li J, Yu H, Zhao Y, Zhang G, Wu Y (2008) Levels of polybrominated diphenyl ethers (PBDEs) in breast milk from Beijing, China. Chemosphere 73:182–186. doi:10.1016/j.chemosphere.2008.05.019

Li J et al (2009) A national survey of polychlorinated dioxins, furans (PCDD/Fs) and dioxin-like polychlorinated biphenyls (dl-PCBs) in human milk in China. Chemosphere 75:1236–1242. doi:10.1016/j.chemosphere.2009.01.073

Lignell S, Aune M, Darnerud PO, Cnattingius S, Glynn A (2009) Persistent organochlorine and organobromine compounds in mother's milk from Sweden 1996-2006: compound-specific temporal trends. Environ Res 109:760–767. doi:10.1016/j.envres.2009.04.011

Lim JS, Lee DH, Jacobs DR Jr (2008) Association of brominated flame retardants with diabetes and metabolic syndrome in the U.S. population, 2003-2004. Diabetes Care 31:1802–1807. doi:10.2337/dc08-0850

Minh NH et al (2004) Persistent organochlorine residues in human breast milk from Hanoi and Hochiminh City, Vietnam: contamination, accumulation kinetics and risk assessment for infants. Environ Pollut 129:431–441. doi:10.1016/j.envpol.2003.11.012

Mishra K, Sharma RC (2011) Assessment of organochlorine pesticides in human milk and risk exposure to infants from North-East India. Sci Total Environ 409:4939–4949. doi:10.1016/j.scitotenv.2011.07.038

Mostafalou S, Abdollahi M (2012) The role of environmental pollution of pesticides in human diabetes. Int J Pharmacol 8:139–140. doi:103923/ijp.2012.139.140

Mostafalou S, Abdollahi M (2013) Pesticides and human chronic diseases: evidences, mechanisms, and perspectives. Toxicol Appl Pharmacol 268:157–177. doi:10.1016/j.taap.2013.01.025

Mostafalou S, Eghbal MA, Nili-Ahmadabadi A, Baeeri M, Abdollahi M (2012) Biochemical evidence on the potential role of organophosphates in hepatic glucose metabolism toward insulin resistance through inflammatory signaling and free radical pathways. Toxicol Ind Health 28:840–851. doi:10.1177/0748233711425073

Mueller JF, Harden F, Toms LM, Symons R, Furst P (2008) Persistent organochlorine pesticides in human milk samples from Australia. Chemosphere 70:712–720. doi:10.1016/j.chemosphere.2007.06.037

Ngwa E, Kengne A-P, Tiedeu-Atogho B, Mofo-Mato E-P, Sobngwi E (2015) Persistent organic pollutants as risk factors for type 2 diabetes. Diabetol Metab Syndr 7:1–15. doi:10.1186/s13098-015-0031-6

Ntow WJ, Tagoe LM, Drechsel P, Kelderman P, Gijzen HJ, Nyarko E (2008) Accumulation of persistent organochlorine contaminants in milk and serum of farmers from Ghana. Environ Res 106:17–26. doi:10.1016/j.envres.2007.05.020

Nyska A, Jokinen MP, Brix AE et al (2004) Exocrine pancreatic pathology in female Harlan Sprague-Dawley rats after chronic treatment with 2,3,7,8-tetrachlorodibenzo-p-dioxin and dioxin-like compounds. Environ Health Perspect 112:903–909, PMCID: PMC1242020

Odegaard JI, Chawla A (2013) Pleiotropic actions of insulin resistance and inflammation in metabolic homeostasis. Science 339:172–177. doi:10.1126/science.1230721

Pal S, Blais JM, Robidoux MA, Haman F, Krummel E, Seabert TA, Imbeault P (2013) The association of type 2 diabetes and insulin resistance/secretion with persistent organic pollutants in two First Nations communities in northern Ontario. Diabetes Metab J 39:497–504. doi:10.1016/j.diabet.2013.01.006

Park SK, Son HK, Lee SK, Kang JH, Chang YS, Jacobs DR, Lee DH (2010) Relationship between serum concentrations of organochlorine pesticides and metabolic syndrome among non-diabetic adults. J Prev Med Public Health 43:1–8. doi:10.3961/jpmph.2010.43.1.1

Park MH et al (2014) Age-related inflammation and insulin resistance: a review of their intricate interdependency. Arch Pharm Res 37:1507–1514. doi:10.1007/s12272-014-0474-6

Persky V et al (2011) Associations of polychlorinated biphenyl exposure and endogenous hormones with diabetes in post-menopausal women previously employed at a capacitor manufacturing plant. Environ Res 111:817–824. doi:10.1016/j.envres.2011.05.012

Persky V et al (2012) Polychlorinated biphenyl exposure, diabetes and endogenous hormones: a cross-sectional study in men previously employed at a capacitor manufacturing plant. Environ Health 11:57. doi:10.1186/1476-069X-11-57

Polder A, Gabrielsen GW, Odland JO, Savinova TN, Tkachev A, Loken KB, Skaare JU (2008a) Spatial and temporal changes of chlorinated pesticides, PCBs, dioxins (PCDDs/PCDFs) and brominated flame retardants in human breast milk from Northern Russia. Sci Total Environ 391:41–54. doi:10.1016/j.scitotenv.2007.10.045

Polder A, Thomsen C, Lindstrom G, Loken KB, Skaare JU (2008b) Levels and temporal trends of chlorinated pesticides, polychlorinated biphenyls and brominated flame retardants in individual human breast milk samples from Northern and Southern Norway. Chemosphere 73:14–23. doi:10.1016/j.chemosphere.2008.06.002

Polder A, Skaare JU, Skjerve E, Loken KB, Eggesbo M (2009) Levels of chlorinated pesticides and polychlorinated biphenyls in Norwegian breast milk (2002-2006), and factors that may predict the level of contamination. Sci Total Environ 407:4584–4590. doi:10.1016/j.scitotenv.2009.04.032

Raab U, Preiss U, Albrecht M, Shahin N, Parlar H, Fromme H (2008) Concentrations of polybrominated diphenyl ethers, organochlorine compounds and nitro musks in mother's milk from Germany (Bavaria). Chemosphere 72:87–94. doi:10.1016/j.chemosphere.2008.01.053

Rignell-Hydbom A, Rylander L, Hagmar L (2007) Exposure to persistent organochlorine pollutants and type 2 diabetes mellitus. Hum Exp Toxicol 26:447–452. doi:10.1177/0960327107076886

Ritter L, Solomon KR, Forget J, Stemeroff M, O'Leary C (2007) Persistent organic pollutants. United Nations Environment Programme

Ruzzin J (2012) Public health concern behind the exposure to persistent organic pollutants and the risk of metabolic diseases. BMC Public Health 12:1–8. doi:10.1186/1471-2458-12-298

Ruzzin J et al (2010) Persistent organic pollutant exposure leads to insulin resistance syndrome. Environ Health Perspect 118:465–471. doi:10.1289/ehp.0901321

Schuhmacher M, Kiviranta H, Vartiainen T, Domingo JL (2007) Concentrations of polychlorinated biphenyls (PCBs) and polybrominated diphenyl ethers (PBDEs) in milk of women from Catalonia, Spain. Chemosphere 67:S295–S300. doi:10.1016/j.chemosphere.2006.05.110

Schuhmacher M, Kiviranta H, Ruokojärvi P, Nadal M, Domingo JL (2009) Concentrations of PCDD/Fs, PCBs and PBDEs in breast milk of women from Catalonia, Spain: a follow-up study. Environ Int 35:607–613. doi:10.1016/j.envint.2008.12.003

She J, Holden A, Sharp M, Tanner M, Williams-Derry C, Hooper K (2007) Polybrominated diphenyl ethers (PBDEs) and polychlorinated biphenyls (PCBs) in breast milk from the Pacific Northwest. Chemosphere 67:S307–S317. doi:10.1016/j.chemosphere.2006.05.154

Someya M et al (2010) Persistent organic pollutants in breast milk of mothers residing around an open dumping site in Kolkata, India: specific dioxin-like PCB levels and fish as a potential source. Environ Int 36:27–35. doi:10.1016/j.envint.2009.09.003

Subramanian A, Ohtake M, Kunisue T, Tanabe S (2007) High levels of organochlorines in mothers' milk from Chennai (Madras) city, India. Chemosphere 68:928–939. doi:10.1016/j.chemosphere.2007.01.041

Sudaryanto A, Kunisue T, Kajiwara N, Iwata H, Adibroto TA, Hartono P, Tanabe S (2006) Specific accumulation of organochlorines in human breast milk from Indonesia: levels, distribution, accumulation kinetics and infant health risk. Environ Pollut 139:107–117. doi:10.1016/j.envpol.2005.04.028

Sudaryanto A, Kajiwara N, Takahashi S, Muawanah, Tanabe S (2008) Geographical distribution and accumulation features of PBDEs in human breast milk from Indonesia. Environ Pollut 151:130–138. doi:10.1016/j.envpol.2007.02.016

Sun SJ et al (2005) Persistent organic pollutants in human milk in women from urban and rural areas in northern China. Environ Res 99:285–293. doi:10.1016/j.envres.2005.05.007

Szyrwińska K, Lulek J (2007) Exposure to specific polychlorinated biphenyls and some chlorinated pesticides via breast milk in Poland. Chemosphere 66:1895–1903. doi:10.1016/j.chemosphere.2006.08.010

Tillett T (2010) Chew on this: persistent organic pollutants may promote insulin resistance syndrome. Environ Health Perspect 118:A173. doi:10.1289/ehp.118-a173b

Todaka T et al (2010) Relationship between the concentrations of polychlorinated dibenzo-p-dioxins, polychlorinated dibenzofurans, and polychlorinated biphenyls in maternal blood and those in breast milk. Chemosphere 78:185–192. doi:10.1016/j.chemosphere.2009.09.047

Toms LM, Harden FA, Symons RK, Burniston D, Furst P, Muller JF (2007) Polybrominated diphenyl ethers (PBDEs) in human milk from Australia. Chemosphere 68:797–803. doi:10.1016/j.chemosphere.2007.02.059

Toms LM, Hearn L, Kennedy K, Harden F, Bartkow M, Temme C, Mueller JF (2009) Concentrations of polybrominated diphenyl ethers (PBDEs) in matched samples of human milk, dust and indoor air. Environ Int 35:864–869. doi:10.1016/j.envint.2009.03.001

Tsang HL, Wu S, Leung CK, Tao S, Wong MH (2011) Body burden of POPs of Hong Kong residents, based on human milk, maternal and cord serum. Environ Int 37:142–151. doi:10.1016/j.envint.2010.08.010

Turyk M, Anderson HA, Knobeloch L, Imm P, Persky VW (2009) Prevalence of diabetes and body burdens of polychlorinated biphenyls, polybrominated diphenyl ethers, and p, p-′-diphenyldichloroethene in Great Lakes sport fish consumers. Chemosphere 75:674–679. doi:10.1016/j.chemosphere.2008.12.035

Uemura H et al (2009) Prevalence of metabolic syndrome associated with body burden levels of dioxin and related compounds among Japan's general population. Environ Health Perspect 117:568–573. doi:10.1289/ehp.0800012

Wahlang B et al (2013) Polychlorinated biphenyl 153 is a diet-dependent obesogen that worsens nonalcoholic fatty liver disease in male C57BL6/J mice. J Nutr Biochem 24:1587–1595. doi:10.1016/j.jnutbio.2013.01.009

Wang YF et al (2008) Associations of polybrominated diphenyl ethers (PBDEs) in breast milk and dietary habits and demographic factors in Taiwan. Food Chem Toxicol 46:1925–1932. doi:10.1016/j.fct.2008.01.032

Wang C, Xu S, Lv Z, Li Y, Wang Y, Chen T (2010a) Exposure to persistent organic pollutants as potential risk factors for developing diabetes. Sci China Chem 53:980–994. doi:10.1007/s11426-010-0157-1

Wang J, Lv X, Du Y (2010b) Inflammatory response and insulin signaling alteration induced by PCB77. J Environ Sci (China) 22:1086–1090

Wieser V, Moschen AR, Tilg H (2013) Inflammation, cytokines and insulin resistance: a clinical perspective. Arch Immunol Ther Exp (Warsz) 61:119–125. doi:10.1007/s00005-012-0210-1

Zhao G, Xu Y, Li W, Han G, Ling B (2007) PCBs and OCPs in human milk and selected foods from Luqiao and Pingqiao in Zhejiang, China. Sci Total Environ 378:281–292. doi:10.1016/j.scitotenv.2007.03.008

Zhou P et al (2011) National survey of the levels of persistent organochlorine pesticides in the breast milk of mothers in China. Environ Pollut 159:524–531. doi:10.1016/j.envpol.2010.10.014

Zietz BP, Hoopmann M, Funcke M, Huppmann R, Suchenwirth R, Gierden E (2008) Long-term biomonitoring of polychlorinated biphenyls and organochlorine pesticides in human milk from mothers living in northern Germany. Int J Hyg Environ Health 211:624–638. doi:10.1016/j.ijheh.2008.04.001

Pharmaceuticals May Disrupt Natural Chemical Information Flows and Species Interactions in Aquatic Systems: Ideas and Perspectives on a Hidden Global Change

Ellen Van Donk, Scott Peacor, Katharina Grosser, Lisette N. De Senerpont Domis, and Miquel Lürling

Contents

E. Van Donk (✉)
Department of Aquatic Ecology, Netherlands Institute of Ecology (NIOO-KNAW), Wageningen, Netherlands

Department of Ecology and Biodiversity, University of Utrecht, Utrecht, Netherlands
e-mail: e.vandonk@nioo.knaw.nl

S. Peacor
Department of Aquatic Ecology, Netherlands Institute of Ecology (NIOO-KNAW), Wageningen, Netherlands

Department of Fisheries and Wildlife, Michigan State University, East Lansing, MI, USA
e-mail: peacor@msu.edu

K. Grosser
Department of Aquatic Ecology, Netherlands Institute of Ecology (NIOO-KNAW), Wageningen, Netherlands

German Centre for Integrative Biodiversity Research (iDiv) Halle-Jena-Leipzig, Deutscher Platz 5e, 04103 Leipzig, Germany
e-mail: katharina.grosser@idiv.de

L.N. De Senerpont Domis • M. Lürling
Department of Aquatic Ecology, Netherlands Institute of Ecology (NIOO-KNAW), Wageningen, Netherlands

Aquatic Ecology & Water Quality Management Group, Dept. Environmental Sciences, Wageningen University, Wageningen, Netherlands
e-mail: l.deSenerpontDomis@nioo.knaw.nl; miquel.lurling@wur.nl

P. de Voogt (ed.), *Reviews of Environmental Contamination and Toxicology, Volume 238*, Reviews of Environmental Contamination and Toxicology 238,
DOI 10.1007/398_2015_5002

1 Introduction

Over the last decades, anthropogenic activities have discharged into the environment many manmade chemicals. There is a rising concern regarding pharmaceutical products and their spread into the environment (e.g. Kümmerer 2008). Due to the enormous quantities consumed, anti-inflammatories, antibiotics, anti-depressives, hormones and blood lipid regulators are found in almost all aquatic environments (Kolpin et al. 2002; Loos et al. 2009). Most pharmaceuticals tend to enter the aquatic environment continuously (but see Sacher et al. 2008 for seasonal exception) in contrast to other pollutants such as herbicides and insecticides which are applied only at specific times related to the life cycle of the target organism, or in response to observed pest outbreaks (Rosi-Marshall and Royer 2012). Pharmaceuticals are designed to be biologically active at very low concentrations and end up in surface waters either unchanged, or as active metabolites/polar conjugates, mostly via municipal wastewater and agricultural discharges (Boxall et al. 2012).

In receiving surface waters, organisms live in a sea of natural chemical substances, released by other organisms, to which many react. Combinations of such chemicals, referred to as infochemicals, constitute a "smellscape" important in shaping and functioning of aquatic ecosystems (see Sect. 2). Pollutants can disrupt these chemically-mediated information flows at several levels in the chemical signaling pathways/networks (reviewed in Lürling and Scheffer 2007; Boyd 2010; Olsén 2011; Lürling 2012). However, one class of emerging pollutants has received far less attention than others: pharmaceuticals. Even at very low concentrations pharmaceuticals may mimic infochemicals or interfere with their operation, due to their structural and functional similarity to the original compounds (Klaschka 2008).

Although direct effects of pharmaceuticals on many organisms have been widely studied, very limited research is devoted to a hidden aspect of pharmaceuticals: do pharmaceuticals affect interactions between species by disrupting infochemical pathways? This review aims to explore ideas and perspectives for the potential impact of pharmaceuticals on the structure and functioning of aquatic ecosystems via disrupting natural chemical information networks between aquatic organisms.

2 Natural Information Transfer via Infochemicals in Aquatic Ecosystems

Organisms use chemical cues in their surroundings, so-called infochemicals, as an important source of information on their environment (Brönmark and Hansson 2012; Vos et al. 2006). Such infochemicals are compounds released by organisms and play a critical role across different organismal functions and interactions, including competition, predation, navigation to and choice of mates, location of resources, and navigation to breeding grounds. For example, infochemical facilitated predator effects on prey phenotypes have been shown to be widespread across diverse taxa, from phytoplankton responding with morphological changes to zooplankton herbivores (Van Donk et al. 2011), to vertebrate predator-prey interactions, across almost every imaginable aquatic ecosystem from streams, ponds, lakes, and marine habitats (reviewed in Brönmark and Hansson 2012; Dodson et al. 1994; Tollrian and Harvell 1999) (Fig. 1).

Lake mesocosm experiments provide evidence of the potential profound role of infochemicals in food web interactions between fish, zooplankton, and phytoplankton (Boeing and Ramcharan 2010). Some of the zooplankton (*Daphnia*) clones responded to fish infochemicals by migrating lower in the water column to avoid predators. The effects of infochemicals cascaded through the food web: without an infochemical induced behavioral response, *Daphnia* were driven to near extinction with associated algal blooms and reduction in fish biomass. On the other hand, with the behavioral response, *Daphnia* population dynamics were stable and fish biomass increased. The ability to assess predation risk via infochemicals goes far

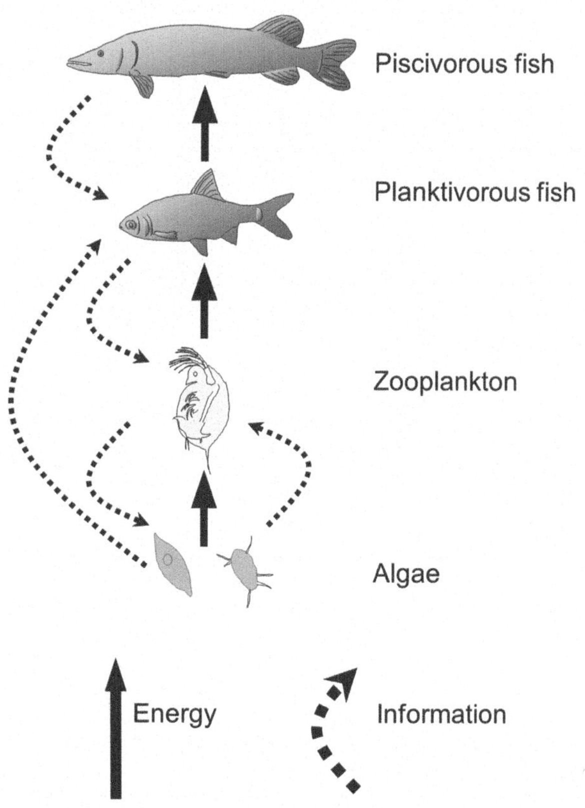

Fig. 1 Typical open water food chain in which energy (*solid black arrow*) flows from algae (primary producers), via zooplankton (herbivores), planktivorous fish (primary consumers) to top predators (piscivorous fish). Within and between the trophic levels chemical cues (infochemicals) convey information (*dotted arrows*). (From Lürling (2012), in Brönmark and Hansson (2012))

beyond simply sensing a single predator's infochemicals. Species have been shown to differentiate the scent of different predators and modulate their response accordingly (Dodson et al. 1994). Furthermore, the ability of a prey species to perceive risk changes with the diet of its predator (Dodson et al. 1994), and some species can even balance risk based on the presence of conspecifics or other prey of the predators (e.g. higher competitor densities represent a weaker predation risk at the same infochemical level, Van Buskirk et al. 2011). In essence, ecologists are discovering that species navigate a complex chemical smell-scape of infochemicals to gauge predation risk, avoid competitors and find food or mates (Fig. 2).

The nature of the chemicals that serve as infochemicals and transfer information between organisms is diverse. It ranges from chemicals that could be considered metabolic products that leak to the environment and fortuitously convey information, to chemicals created by organisms to serve particular

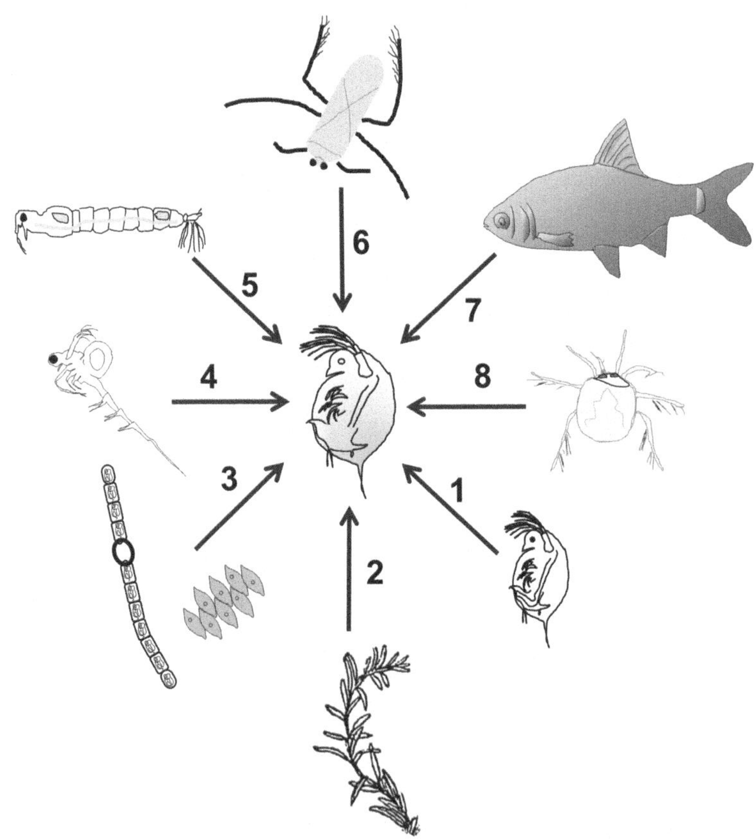

Fig. 2 Examples of a smell-scape in which the central crustacean zooplankton *Daphnia* receives chemical information from conspecifics (*1*), plants (*2*), phytoplankton food (*3*), zooplankton predators (*4*), insect predators (*5, 6*), vertebrate predators (*7*) and arachnid predators (*8*). See Tollrian and Harvell (1999) and Brönmark and Hansson (2012)

purposes such as alarm signals (Dodson et al. 1994). Chemists have identified a large number of such substances involved in interactions among organisms in terrestrial and aquatic systems (e.g. Pohnert et al. 2007). Many of the infochemicals in aquatic ecosystems occur at very low concentrations. Over evolutionary time the chemosensory systems have evolved to be finely tuned to detect and react to these compounds.

3 Current Levels of Pharmaceuticals Affect Organisms

3.1 Direct and Indirect Effects

Recent studies on the effects of pharmaceuticals show that they can have profound effects on organisms at levels found in natural systems. At the most extreme, pharmaceutical compounds have been found to lead directly to the death of individuals. Three species of vulture in Asia have been brought to the edge of extinction due to renal failure following consumption of carcasses of diclofenac-treated livestock (Oaks et al. 2004). If animals are preyed upon such direct effects may not only affect the prey, but travel further through the food web via indirect effects on species that interact with the prey (Relyea and Hoverman 2006).

In aquatic environments, pharmaceutical concentrations are in general orders of magnitude lower than the concentrations exerting direct lethal effects in laboratory assays (Santos et al. 2010). Nonetheless, a growing number of laboratory studies report density-mediated direct effects at realistic concentrations. For instance, tetracycline concentrations as low as 0.5 μg/L led to lower bacteria and cyanobacteria biomass in periphyton of artificial streams (Quinlan et al. 2011); a mixture of low concentrations 17α-ethynylestradiol (10 ng/L) and fluoxetine (10 ng/L) significantly reduced population growth rates for *Physa pomilia* snails (Luna et al. 2013); a 21 day exposure of adult male fathead minnows (*Pimephales promelas*) to environmentally realistic concentrations of sertraline (5.2 ng/L) and venlafaxine (305 ng/L) resulted in mortality (Schultz et al. 2011). Other reports show direct effects on endpoints that likely correlate with growth rate or mortality: clotrimazole is found in low concentrations (ng/L) in natural systems and a similar concentration (17 ng/L) caused inhibition of algal 14α-demethylase in lab experiments (OSPAR Commission 2013); exposure to diclofenac (1 μg/L) caused structural disruptions in the kidney and intestine of rainbow trout (Mehinto et al. 2010).

Importantly, direct effects may only become apparent after prolonged exposure, potentially through multigenerational effects. While standard acute and chronic assays with the zooplankton grazer *Ceriodaphnia* indicated that toxic effects of sertraline at environmentally relevant concentrations were unlikely, a simple extension of the experimental duration showed that in the third generation effects on growth and reproduction occurred at a concentration of 4.8 μg/L, which is only a few times higher than levels that have been encountered in nature (Lamichhane et al. 2014). Growth in fathead minnows was reduced after 58 days exposure to 4 ng/L 17α-ethynylestradiol in the F_0 parent population, but growth reduction occurred already after 28 days at 0.2 ng/L in the offspring F_1 population (Länge et al. 2001). Exposure of fathead minnow during 3 years to environmental realistic concentrations of 17α-ethynylestradiol (4.8–6.1 ng/L), led to a complete collapse of the population in the treated lake (Kidd et al. 2007). These examples illustrate that continuous exposure to a pharmaceutical during multiple generations may not only lead to increased sensitivity over time (Länge et al. 2001; Lamichhane et al. 2014), but also that long term exposure may have an impact on the whole population (Kidd et al. 2007).

Higher mortality and reduced population growth of more sensitive species (S1 in Fig. 3a) may also influence more tolerant or insensitive species (S2 in Fig. 3a) via indirect effects (Fleeger et al. 2003). In fact, these density-mediated indirect effects caused by pollutants might be quite common (Relyea and Diecks 2008). However, the effect of one species on another might not just travel through densities, but species interactions can also be influenced by changes in activities, behaviour or phenotypic traits (Fig. 3b). Low concentrations of pollutants and pharmaceuticals

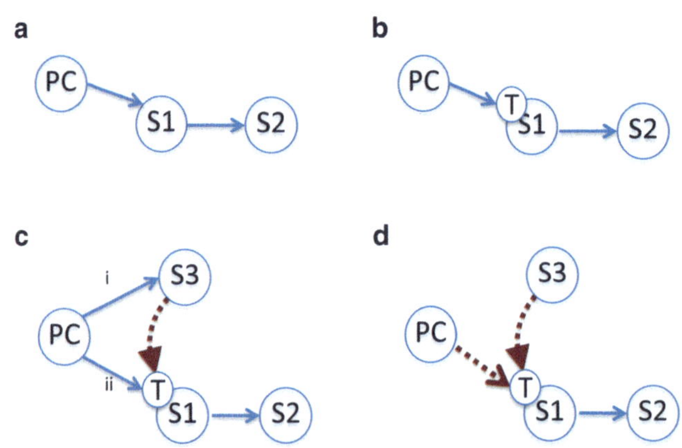

Fig. 3 Multiple pathways by which pharmaceutical compounds may directly and indirect affect species. In all cases the fitness of the focal species (S1) may be affected, which in turn can indirectly affect other species (S2) in the system that interact with S1, including resources, competitors and other predators. (**a**) Density mediated effects: pharmaceuticals may have a toxic effect killing or lowering the density of focal species S1 through reduced growth. (**b**) Trait-mediated effects: pharmaceuticals may affect traits (T) of the focal species by affecting e.g. the nervous system or having sublethal effects on the condition of the focal species. (**c**) Infochemical disruption: Pharmaceuticals affect the transfer of information from a sender species, S3, to a receiver species S1, by disrupting (*i*) the senders' production of infochemicals, or (*ii*) the receivers' reception of the infochemical. (**d**) Infochemical mimicry: Pharmaceuticals may mimic infochemicals, causing changes in trait expression of S1. For simplicity we label the nodes in the diagrams "species", however they can also represent other groups, such as different sexes of the same species

encountered in nature, rather than primarily leading to effects on mortality, may also show changes in species' traits (Fig. 3b). There is now convincing evidence that pharmaceuticals may cause feminization of male fish at estrogen concentrations in the lower ng/L range (Gross-Sorokin et al. 2006). Behavioral changes of fish at environmental concentrations of 1.8 μg/L oxazepam (Brodin et al. 2013), 0.12 μg/L sertraline (Hedgespeth et al. 2014) and ~1.1 μg/L oxazepam (Klaminder et al. 2014) have been reported. In the amphipod *Gammarus pulex* exposure to environmentally realistic concentrations of ibuprofen (10 ng/L) or fluoxetine (100 ng/L) led to decreased activity (De Lange et al. 2006), whereas ibuprofen (1, 10 and 100 ng/L), fluoxetine (10 and 100 ng/L) and carbamazepine (1 and 10 ng/L) caused elevated ventilation (De Lange et al. 2009). As with direct effects on density, direct effects of pharmaceuticals on a species' traits could also indirectly affect other species (Fig. 3b, S1 → S2) through modification of the interaction strengths between the affected species with the other species. This is analogous to the case, for example, when predators cause (through induction) a change in prey traits that leads to indirect effects of the predator on resources, competitors and other predators of the prey (Werner and Peacor 2003).

3.2 Infodisruption and Mimicry

The fact that pharmaceutical concentrations occur at levels in the natural environment high enough to affect species traits is a foreboder that these levels may affect infochemical pathways as well (Fig. 3c, d). We divide the potential effects on species interactions into two broad categories. First, (Fig. 3c) pharmaceuticals may disrupts the transfer of information by the infochemical either by (a) affecting the production of the infochemical by the sender (Fig. 3c, arrow i), or (b) affecting the reception of the infochemical by the receiver (Fig. 3c, arrow ii). Second, pharmaceuticals may mimic infochemicals (Fig. 3d). Both disruption (Fig. 3c) and mimicry (Fig. 3d) could affect the receiver species and indirectly affect other species by affecting the receiver species traits.

An analysis of published literature shows a strong increase in the amount of publications on pharmaceuticals and infochemicals since 1990 (Fig. 4). Despite this large body of literature, only a small number of publications cover both infochemicals and pharmaceuticals. Although this subject is little investigated, we review below the studies that point to the potential role pharmaceuticals could play.

The majority of studies of pharmaceutical effects on infochemical-mediated interactions have examined a pathway in which pharmaceuticals affect the reception (i.e. perception and processing) of infochemicals by the receiver (Fig. 3c-ii). For example, an antidepressant, fluoxetine, has been shown in laboratory studies to interfere with reception in a number of fish species by (a) disrupting the integration of pheromone cues to control sexual behaviors in male gold fish (Mennigen et al. 2010—54 μg/L), (b) causing elevated alarm responses in Arabian killifish (Barry 2013—0.03–3 μg/L) and (c) slowing predator avoidance response in larval

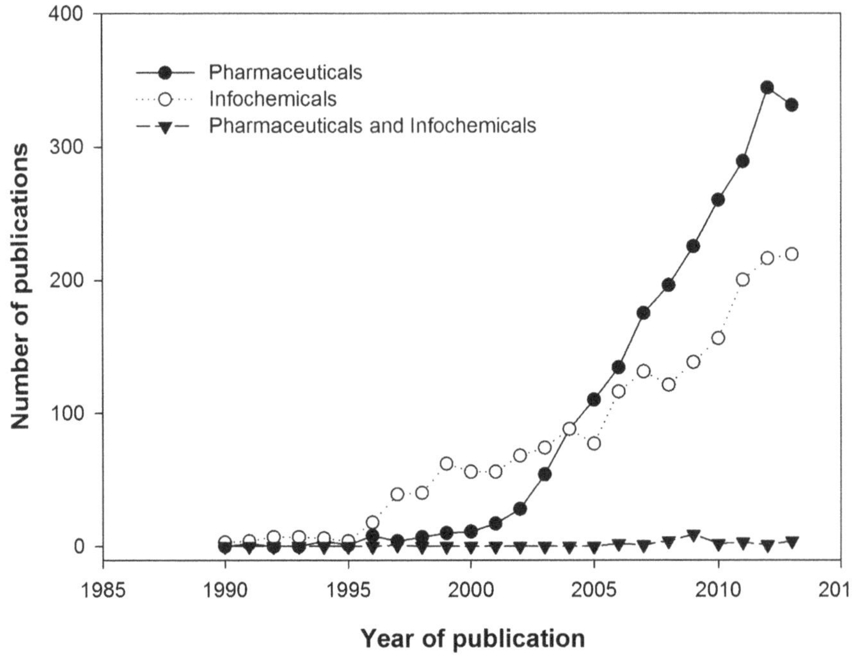

Fig. 4 Analysis of published literature based on combinations of the keywords "pharmaceutical", "infochemicals", "chemical cues" (the latter three summarized in the graph as infochemicals) with either "aquatic", "marine" or "freshwater" in different databases (Web of Science and Scopus). Excluded from analyses is literature on natural products which could potentially serve as pharmaceutical

fathead minnows (Painter et al. 2009—25 ng/L). Other effects have been found in vastly different pharmaceutical compounds, including (a) propranolol lowering the response of amphipods to predator cues, albeit at rather high concentrations of 100–5000 μg/L (Wiklund et al. 2011), (b) the painkiller acetylsalicylic acid (1 mM) impairing the larval metamorphosis of the large sea snail queen conch (*Strombus gigas*) that is triggered by red algal (*Laurencia poitei*) chemical cues (Boettcher and Target 1998), and (c) the veterinary pharmaceutical ivermectin (at 10 mg/L)—a broad-spectrum antiparasitic agent—blocking the chemoreception of allelochemicals and pheromones in nematodes (Rolfe et al. 2001). There is also circumstantial evidence how pharmaceuticals could interfere with the reception of infochemicals (Fig. 3c-ii). For example, the antidepressant fluoxetine affected amphipod photo- and geotaxis at 0.1 μg/L and higher levels, which could impair anti-predator behavior (Guler and Ford 2010).

There is much less evidence for the potential influence of pharmaceuticals on the production of infochemicals by the sender (Fig. 3c-i). We are only aware of studies using fish as a model organism, where hormonal steroids that are released into the water can act as potent pheromones (Stacey 2011). In female goldfish during vitellogenesis (yolk incorporation in the oocytes), the hormone 17β-estradiol stimulates urinary release of an unidentified pheromone that attracts males, while the oocyte maturation-inducing steroid 17,20β-dihydroxy-4-pregnen-3-one (17,20β-P) acts as a pheromone affecting male hormone levels and sexual behaviors (Stacey 2011). The antidepressant fluoxetine reduces the level of 17β-estradiol in female goldfish (Mennigen et al. 2008), whereas the oral contraceptive levonorgestrel reduces 17,20β-P in female fathead minnows (Overturf et al. 2012), and thus effects on pheromone communication are likely. A simplified working model on how the neuroendocrine disruption in fish by fluoxetine might affect pheromone communication via sex steroids is visualized in Fig. 5 in Mennigen et al. (2011). In addition, gestagens (natural progestogens and synthetic progestins) have been identified as class of pharmaceuticals that need to be studied in relation to potential effects on pheromonal communication (Orlando and Ellestad 2014).

A number of studies suggest that pharmaceuticals may mimic infochemicals, representing another mechanism by which infochemicals disrupt interactions (Fig. 3d). For instance, several androgens and progestins—that were detected in effluent at concentrations up to 14.9 ng/L, often exceeding olfactory detection thresholds for pheromones in fish—are expected to disrupt pheromone communication in fish either through eliciting responses at inappropriate times or through competitive binding to olfactory receptors (Kolodziej et al. 2003). In a second example, the antidepressants venlafaxine and citalopram caused foot detachment in freshwater snails at environmentally realistic concentrations as low as 313 pg/L and 405 pg/L, respectively (Fong and Hoy 2012), which in marine snails is a known chemical stimuli-mediated escape response to predatory starfish (Lemmnitz et al. 1989). Examples of fluoxetine-induced spawning of freshwater mussels (Bringolf et al. 2010)—even at low fluoxetine concentrations of 50 nM (Fong 1998) and 20 ng/L (Lazzara et al. 2012)—might point at mimicry, as during mass spawning events, mussels use sex pheromones for attracting the sexual partner and coordinated release of gametes by both partners (Paul et al. 2011). Pharmaceuticals mediated spawning under unfavorable environmental conditions can potentially reduce mussel reproduction and ultimately lead to a change in the trophic structure.

These examples underscore that pharmaceuticals can impact infochemical-mediated interactions and might thereby affect the fitness of the organisms involved and potentially food-web structure. There are presently few studies in this area. Indeed, we are aware of no studies that have investigated potential ensuing indirect effects on other species (i.e. S2 in Fig. 3c, d) and in turn the larger ecological community. To our knowledge, all published effects of pharmaceuticals on species interactions via infochemical pathways are examples in which the sender and/or receiver species are affected. Presumably such direct effects will propagate to indirect effects on other species via the pathways outlined in Fig. 3c, d. For instance, the selective serotonin reuptake inhibitor sertraline—impaired feeding of perch on the zooplankton grazer *Daphnia* in a concentration range between 0.12 and 89 μg/L (Hedgespeth et al. 2014). In contrast, Brodin et al. (2013) found that low concentrations (1.8 μg/L) of the anxiolytic drug oxazepam increased feeding activity of perch on *Daphnia*. Likewise, Klaminder et al. (2014) found increased activity and lower mortality rates in perch (*Perca fluviatilis*) exposed to 1.1/1.2 μg/L oxazepam.

In latter two studies, perch showed besides increased feeding activity on *Daphnia*, also less sociality and more bold behaviour (Brodin et al. 2013; Klaminder et al. 2014). Although bold individuals tend to grow faster, such behavioural changes may lead to higher risk of predation (Hellström and Magnhagen 2011) and thus the outcome of behavioural changes will depend on the environmental context. Either way, a stronger or reduced predation pressure of perch on the grazer *Daphnia* will influence the information flow between these organisms that will travel further to phytoplankton (Ringelberg 2009) which may have consequences for other trophic groups as well (Lürling and Van Donk 1997). We believe further research in this area will uncover such indirect effects because the perch-*Daphnia*-algae tritrophic food chain represents well known infochemical pathways and as illustrated by Ringelberg (2009) the information network can be viewed as superimposed on and tightly connected to the flow of matter (see Fig. 13.2 in Ringelberg 2009). Therefore, it is highly likely that effects on perch will not only influence the flow of matter, but also the superimposed information network. Such model systems will make good candidates to explore the role of pharmaceuticals, as effects on the information flow (infochemical network) can be separated from effects on the energy flow using existing, well-developed bioassays.

4 Conclusions and Future Directions

Many aquatic organisms use infochemicals not only to find partners and food, but also to sense the presence of natural enemies and to avoid predation. Although there is a growing body of evidence that a wide variety of anthropogenic pollutants commonly found in surface waters—at environmentally realistic concentrations—can impair chemical communication between aquatic organisms, the impact of pharmaceuticals has received far less attention than other pollutants. Our review indicates that at very low concentrations pharmaceuticals may mimic infochemicals or interfere with their operation, due to their structural and functional similarity to the original compounds. So these biologically active pharmaceuticals may pose a risk of disruption of the ubiquitous natural chemical information transfer between organisms. Combined with a plethora of potential other stressors influencing their mode of action, this makes pharmaceuticals a true hidden global change.

In the future several major challenges need to be addressed to further substantiate the incidence and scale at which infodisruption takes place. Virtually all studies refer to laboratory experiments with single pollutants, while in their natural environment organisms are potentially confronted with multiple infodisruptors acting in concert under varying conditions. Importantly, to understand the impact of such infodisruption on natural populations and ecosystems, multi-species and multi-trophic experiments in mesocosms, combined with multi-compound exposures and model studies are needed to advance the field. Our review underlines that effects of pharmaceuticals go beyond common practice endpoints, we therefore would like to promote the initiative to extend current ecotoxicological testing of pharmaceuticals (Klaminder et al. 2014; Brodin et al. 2014). Use of standard, well-known model systems, such as the fish-*Daphnia* (Ringelberg 2009) and *Daphnia-Scenedesmus* systems (Lürling and Van Donk 1997), as well as benthic systems using *Gammarus* sp. (De Lange et al. 2005, 2006, 2009) would be particularly useful in this new generation of exotoxicological experiments.

Another step forward would be the prolonged exposure of above-mentioned model systems as well as more complex communities to blends of low concentrations of pharmaceuticals or even mixed with other pollutants or stressors. A recent study showed that the chemosensory perception of predators by the gray tree frog was reduced by 50 % when tadpoles were housed in polluted stream water and wastewater effluent compared to clean tap water (Troyer and Turner 2015). The substances identified to have an info-disrupting effect have often been hit upon by chance. A systematic scan of selected chemicals and natural-polluting mixes as discharged from wastewater treatments should provide a broader image of the problem.

The potential impact of concentrations found in the environment recently led the EU to include three pharmaceuticals, i.e. 17α-ethynylestradiol, 17β-estradiol and diclofenac, on a watch list for priority substances presenting a significant risk to or via the aquatic environment (EC 2012). This EU watch-list mechanism has been set up for targeted monitoring of potential harmful pharmaceuticals to be included in future lists of priority substances. For cost reasons, however, the list only focuses on a limited number of potential substances in a limited number of monitoring sites in EU countries. Prioritization of pharmaceuticals has been recognized as one of the big questions in the field of pharmaceuticals in the environment (Boxall et al. 2012), yet the sheer number of compounds calls for novel prioritization schemes (e.g. De Voogt et al. 2009; Caldwell et al. 2014). Recently, Caldwell et al. (2014) proposed an intelligent testing strategy to identify what further studies and data are needed to advance the prioritization process.

At present, we have very little knowledge on the extent pharmaceuticals are concentrated in the system or broken down, and if biomagnification takes place. More insight is needed into the fate and trophic transfer of pharmaceuticals in surface waters (Heberer 2002). We encourage recent efforts to study the relevance of metabolites (e.g. Klaminder et al. 2015. Finally, in the near future, the need for clean, healthy water of our expanding world population is an increasingly recognized international challenge, making innovations in waste water treatment (Jung et al. 2015) and production of more biodegradable pharmaceuticals a top priority (Rastogi et al. 2014).

5 Summary

Pharmaceuticals consumption by humans and animals is increasing substantially, leading to unprecedented levels of these compounds in aquatic environments worldwide. Recent findings that concentrations reach levels that can directly have negative effects on organisms are important per se, but also sound an alarm for other potentially more pervasive effects that arise from the interconnected nature of ecological communities. Aquatic organisms use chemical cues to navigate numerous challenges, including the location of mates and food, and the avoidance of natural enemies. Low concentrations of pharmaceuticals can disrupt this "smellscape" of information leading to maladaptive responses. Furthermore, direct effects of pharmaceuticals on the traits and abundance of one species can cascade through a community, indirectly affecting other species. We review mechanisms by which pharmaceuticals in surface waters can disrupt natural chemical information flows and species interactions. Pharmaceuticals form a new class of chemical threats, which could have far-reaching implications for ecosystem functioning and conservation management.

Acknowledgements We want to thank Wim van der Putten and two anonymous reviewers for constructive comments on an earlier draft of the manuscript. This is publication 5932 of the Netherlands Institute of Ecology (NIOO-KNAW).

References

Barry MJ (2013) Effects of fluoxetine on the swimming and behavioural responses of the Arabian killifish. Ecotoxicology 22:425–432

Boeing WJ, Ramcharan CW (2010) Inducible defences are a stabilizing factor for predator and prey populations: a field experiment. Freshw Biol 55:2332–2338

Boettcher AA, Target NM (1998) Role of chemical inducers in larval metamorphosis of queen conch, *Strombus gigas* Linnaeus: relationship to other marine invertebrate systems. Biol Bull 194:132–142

Boxall ABA, Rudd MA, Brooks BW et al (2012) Pharmaceuticals and personal care products in the environment: what are the big questions? Environ Health Perspect 120:1221–1229

Boyd RS (2010) Heavy metal pollutants and chemical ecology: exploring new frontiers. J Chem Ecol 36:46–58

Bringolf RB, Heltsley RM, Newton TJ et al (2010) Environmental occurrence and reproductive effects of the pharmaceutical fluoxetine in native freshwater mussels. Environ Toxicol Chem 29:1311–1318

Brodin T, Fick J, Jonsson M et al (2013) Dilute concentrations of a psychiatric drug alter behavior of fish from natural populations. Science 339:814–815

Brodin T, Piovano S, Fick J, Klaminder J, Heynen M, Jonsson M (2014) Ecological effects of pharmaceuticals in aquatic systems-impacts through behavioural alterations. Philos Trans R Soc Lond B Biol Sci 369:20130580

Brönmark C, Hansson L-A (2012) Aquatic chemical ecology: new directions and challenges for the future. In: Brönmark C, Hansson L-A (eds) Chemical ecology in aquatic systems. Oxford University Press, New York, NY

Caldwell DJ, Mastrocco F, Margiotta-Casaluci L, Brooks BW (2014) An integrated approach for prioritizing pharmaceuticals found in the environment for risk assessment, monitoring and advanced research. Chemosphere 115:4–12

De Lange HJ, Lürling M, Van Den Borne B, Peeters ETHM (2005) Attraction of the amphipod *Gammarus pulex* to water-borne cues of food. Hydrobiologia 544:19–25

De Lange HJ, Noordhoven W, Murk AJ, Lürling M, Peeters ETHM (2006) Behavioural responses of *Gammarus pulex* (Crustacea, Amphipoda) to low concentrations of pharmaceuticals. Aquat Toxicol 78:209–216

De Lange HJ, Peeters ETHM, Lürling M (2009) Changes in ventilation and locomotion of *Gammarus pulex* (Crustacea, Amphipoda) in response to low concentrations of pharmaceuticals. Hum Ecol Risk Assess 15(1):111–120

De Voogt P, Janex-Habibi ML, Sacher F, Puijker L, Mons M (2009) Development of a common priority list of pharmaceuticals relevant for the water cycle. Water Sci Technol 59:39–46

Dodson SI, Crowl TA, Peckarsky BL et al (1994) Non-visual communication in freshwater benthos - an overview. J N Am Benthol Soc 13:268–282

EC. 2012. Directive of the European Parliament and of the Council amending Directives 2000/60/EC and 2008/105/EC as regards priority substances in the field of water policy. In: 2011/0429 (COD). http://eur-lex.europa.eu/legal-content/EN/NOT/?uri=CELEX:52011PC0876

Fleeger JW, Carmana KR, Nisbet RM (2003) Indirect effects of contaminants in aquatic ecosystems. Sci Total Environ 317:207–233

Fong PP (1998) Zebra mussel spawning is induced in low concentrations of putative serotonin reuptake inhibitors. Biol Bull 194:143–149

Fong PP, Hoy CM (2012) Antidepressants (venlafaxine and citalopram) cause foot detachment from the substrate in freshwater snails at environmentally relevant concentrations. Mar Environ Res 45:145–153

Gross-Sorokin MY, Roast SD, Brighty GC (2006) Assessment of feminization of male fish in English rivers by the Environment Agency of England and Wales. Environ Health Perspect 114 (suppl 1):147–151

Guler Y, Ford AT (2010) Anti-depressants make amphipods see the light. Aquat Toxicol 99:397–404

Heberer T (2002) Occurrence, fate, and removal of pharmaceutical residues in the aquatic environment: a review of recent research data. Toxicol Lett 131(1-2):5–17

Hedgespeth ML, Nilsson PA, Berglund O (2014) Ecological implications of altered fish foraging after exposure to an antidepressant pharmaceutical. Aquat Toxicol 151:84–87

Hellström G, Magnhagen C (2011) The influence of experience on risk taking: results from a common-garden experiment on populations of Eurasian perch. Behav Ecol Sociobiol 65:1917–1926

Jung C, Son A, Her N, Zoh KD, Cho J, Yoon Y (2015) Removal of endocrine disrupting compounds, pharmaceuticals, and personal care products in water using carbon nanotubes: a review. J Ind Eng Chem 27:1–11

Kidd KA, Blanchfield PJ, Mills KH, Palace VP, Evans RE, Lazorchak JM, Flick RW (2007) Collapse of a fish population after exposure to a synthetic estrogen. Proc Natl Acad Sci U S A 104:8897–8901

Klaminder J, Jonsson M, Fick J, Sundelin A, Brodin T (2014) The conceptual imperfection of aquatic risk assessment tests: highlighting the need for tests designed to detect therapeutic effects of pharmaceutical contaminants. Environ Res Lett 9(8):084003

Klaminder J, Brodin T, Sundelin A, Anderson NJ, Fahlman J, Jonsson M, Fick J (2015) Long-term persistence of an anxiolytic drug (oxazepam) in a large freshwater lake. Environ Sci Tech 49 (17):10406–10412

Klaschka U (2008) The infochemical effect—a new chapter in ecotoxicology. Environ Sci Pollut Res 15:452–462

Kolodziej EP, Gray JL, Sedlak DL (2003) Quantification of steroid hormones with pheromonal properties in municipal wastewater effluent. Environ Toxicol Chem 22:2622–2629

Kolpin DW, Furlong ET, Meyer MT et al (2002) Pharmaceuticals, hormones, and other organic wastewater contaminants in U.S. streams, 1999–2000: a National reconnaissance. Environ Sci Tech 36:1202–1211

Kümmerer K (2008) Pharmaceuticals in the environment. Sources, fate, effects and risks. Springer, Berlin

Lamichhane K, Garcia SN, Huggett DB et al (2014) Exposures to a selective serotonin reuptake inhibitor (SSRI), sertraline hydrochloride, over multiple generations: changes in life history traits in *Ceriodaphnia dubia*. Ecotoxicol Environ Saf 101:124–130

Länge R, Hutchinson TH, Croudace CP, Siegmund F, Schweinfurth H, Hampe P, Panter GH, Sumpter JP (2001) Effects of the synthetic estrogen 17α-ethinylestradiol on the life-cycle of the fathead minnow (*Pimephales promelas*). Environ Toxicol Chem 20(6):1216–1227

Lazzara R, Blazquez M, Porte C, Barata C (2012) Low environmental levels of fluoxetine induce spawning and changes in endogenous estradiol levels in the zebra mussel *Dreissena polymorpha*. Aquat Toxicol 106:123–130

Lemmnitz G, Schuppe H, Wolff HG (1989) Neuromotor bases of the escape behavior of *Nassa mutabilis*. J Exp Biol 143:493–507

Loos R, Gawlik BM, Locoro G et al (2009) EU-wide survey of polar organic persistent pollutants in European river waters. Environ Pollut 157:561–568

Luna TO, Plautz SC, Salice CJ (2013) Effects of 17α-ethynylestradiol, fluoxetine, and the mixture on life history traits and population growth rates in a freshwater gastropod. Environ Toxicol Chem 32(12):2771–2778

Lürling M (2012) Infodisruption: pollutants interfering with the natural chemical information conveyance in aquatic systems. In: Brönmark C, Hansson L-A (eds) Chemical ecology in aquatic systems. Oxford University Press, New York, NY, pp 250–271

Lürling M, Scheffer M (2007) Info-disruption: pollution and the transfer of chemical information between organisms. Trends Ecol Evol 22:374–379

Lürling M, Van Donk E (1997) Morphological changes in *Scenedesmus* induced by infochemicals released in situ from zooplankton grazers. Limnol Oceanogr 42:783–788

Mehinto AC, Hill EM, Tyler CR (2010) Uptake and biological effects of environmentally relevant concentrations of the nonsteroidal anti-inflammatory pharmaceutical diclofenac in rainbow trout (*Oncorhynchus mykiss*). Environ Sci Tech 44:2176–2182

Mennigen JA, Martyniuk CJ, Crump K et al (2008) Effects of fluoxetine on the reproductive axis of female goldfish (*Carassius auratus*). Physiol Genomics 35:273–282

Mennigen JA, Lado WE, Zamora JM et al (2010) Waterborne fluoxetine disrupts the reproductive axis in sexually mature male goldfish, *Carassius auratus*. Aquat Toxicol 100:354–364

Mennigen JA, Stroud P, Zamora JM, Moon TW, Trudeau VL (2011) Pharmaceuticals as neuroendocrine disruptors: lessons learned from fish on Prozac. J Toxicol Environ Health B Crit Rev 14(5-7):387–412

Oaks JL, Gilbert M, Virani MZ et al (2004) Diclofenac residues as the cause of vulture population decline in Pakistan. Nature 427:630–633

Olsén KH (2011) Effects of pollutants on olfactory mediated behaviors in fish and crustaceans. In: Breithaupt T, Thiel M (eds) Chemical communication in crustaceans. Springer Science + Business Media LLC, New York, NY, pp 507–529, Doi: 10.1007/978-0-387-77101-4-26

Orlando EF, Ellestad LE (2014) Sources, concentrations, and exposure effects of environmental gestagens on fish and other aquatic wildlife, with an emphasis on reproduction. Gen Comp Endocrinol 203:241–249

OSPAR Commission (2013) Background Document on Clotrimazole (2013 update). Publication Number: 595/2013. ISBN 978-1-909159-28-0.

Overturf MD, Overturf CL, Baxter D et al (2012) Early life-stage toxicity of eight pharmaceuticals to the fathead minnow, *Pimephales promelas*. Arch Environ Contam Toxicol 62:455–464

Painter MM, Buerkley MA, Julius ML et al (2009) Antidepressants at environmentally relevant concentrations affect predator avoidance behavior of larval fathead minnows (*Pimephales promelas*). Environ Toxicol Chem 28:2677–2684

Paul VJ, Ritson-Williams R, Sharp K (2011) Marine chemical ecology in benthic environments. Nat Prod Rep 28:345–387

Pohnert G, Steinke M, Tollrian R (2007) Chemical cues, defence metabolites and the shaping of pelagic interspecific interactions. Trends Ecol Evol 22:198–204

Quinlan EL, Nietch CT, Blocksom K, Lazorchak JM, Batt AL, Griffiths R, Klemm DJ (2011) Temporal dynamics of periphyton exposed to tetracycline in stream mesocosms. Environ Sci Tech 45:10684–10690

Rastogi T, Leder C, Kummerer K (2014) Designing green derivatives of beta-blocker metoprolol: a tiered approach for green and sustainable pharmacy and chemistry. Chemosphere 111:493–499

Relyea RA, Diecks N (2008) An unforeseen chain of events: lethal effects of pesticides on frogs at sublethal concentrations. Ecol Appl 18:1728–1742

Relyea R, Hoverman J (2006) Assessing the ecology in ecotoxicology: a review and synthesis in freshwater systems. Ecol Lett 9:1157–1171

Ringelberg J (2009) Diel vertical migration of zooplankton in lakes and oceans: causal explanations and adaptive significances. Springer, New York, NY. ISBN ISBN 978-90-481-3092-4

Rolfe R, Barrett J, Perry R (2001) Electrophysiological analysis of responses of adult females of *Brugia pahangi* to some chemicals. Parasitology 122:347–357

Rosi-Marshall EJ, Royer TV (2012) Pharmaceutical compounds and ecosystem function: an emerging research challenge for aquatic ecologists. Ecosystems 15:867–880

Sacher F, Ehmann M, Gabriel S, Graf C, Brauch HJ (2008) Pharmaceutical residues in the river Rhine - results of a one-decade monitoring programme. J Environ Monit 10:664–670

Santos LHMLM, Araújo AN, Fachini A et al (2010) Ecotoxicological aspects related to the presence of pharmaceuticals in the aquatic environment. J Hazard Mater 175:45–95

Schultz MM, Painter MM, Bartell SE, Logue A, Furlong ET, Werner SL, Schoenfuss HL (2011) Selective uptake and biological consequences of environmentally relevant antidepressant pharmaceutical exposures on male fathead minnows. Aquat Toxicol 104:38–47

Stacey N (2011) Hormonally derived sex pheromones in fishes. In: Norris OD, Lopez KH (eds) Hormones and reproduction of vertebrates. Elsevier, Atlanta, GA

Tollrian R, Harvell D (1999) The ecology and evolution of inducible defenses. The ecology and evolution of inducible defenses. Princeton University Press, New York, NY

Troyer RR, Turner AM (2015) Chemosensory perception of predators by larval amphibians depends on water quality. PLoS One 10(6):e0131516

Van Buskirk J, Ferrari M, Kueng D, Napflin K, Ritter N (2011) Prey risk assessment depends on conspecific density. Oikos 120:1235–1239

Van Donk E, Ianora A, Vos M (2011) Induced defences in marine and freshwater phytoplankton: a review. Hydrobiologia 668:3–19

Vos M, Vet LEM, Wackers FL et al (2006) Infochemicals structure marine, terrestrial and freshwater food webs: implications for ecological informatics. Ecol Inform 1:23–32

Werner EE, Peacor SD (2003) A review of trait-mediated indirect interactions in ecological communities. Ecology 84:1083–1100

Wiklund AE, Oskarsson H, Thorsn G et al (2011) Behavioural and physiological responses to pharmaceutical exposure in macroalgae and grazers from a Baltic Sea littoral community. Aquat Biol 14:29–39

Technologically Enhanced Naturally Occurring Radioactive Materials (TENORM) in the Oil and Gas Industry: A Review

Israel Doyi, David Kofi Essumang, Samuel Dampare, and Eric Tetteh Glover

Contents

1 Introduction

Radiation is part of the natural environment: it is estimated that approximately 80 % of all human exposure comes from naturally occurring or background radiation. Certain extractive industries such as mining and oil logging have the potential to increase the risk of radiation exposure to the environment and humans by

I. Doyi (✉) • S. Dampare
National Radioactive Waste Management Centre, National Nuclear Research Institute, Ghana Atomic Energy Commission, Accra, Ghana
e-mail: i.doyi@gaecgh.org; dampee_2000@yahoo.com

D.K. Essumang
Graduate School of Nuclear and Allied Sciences, Ghana Atomic Energy Commission, Accra, Ghana
e-mail: kofiessumang@yahoo.com; dessumang@ucc.edu.gh

E.T. Glover
Department of Chemistry, University of Cape Coast, Cape Coast, Ghana
e-mail: e.glover@gaecgh.org

P. de Voogt (ed.), *Reviews of Environmental Contamination and Toxicology, Volume 238*, Reviews of Environmental Contamination and Toxicology 238,
DOI 10.1007/398_2015_5005

concentrating the quantities of naturally occurring radiation beyond normal background levels (Azeri-Chirag-Gunashli 2004).

Doyi et al. (2013) reported radiation doses to miners due to airborne radon concentration ranging from 0.06 ± 0.02 to 1.23 ± 0.04 mSvy^{-1}. Naturally occurring radioactive materials (NORMs) in gold ore samples from the artisanal gold mines were also studied and the annual effective doses to the underground miners ranged from 0.11 ± 0.02 to 0.69 ± 0.04 mSvy^{-1}.

Small, but measurable concentrations of NORMs primarily from most minerals are from the ^{238}U and ^{232}Th decay series and ^{40}K (Snavely 1989). It has even been suggested that the long term emission of alpha particles from natural radionuclides could be one of the possible sources of energy associated with the transformation of organic matter into petroleum (Pasocha 1997). In the exploration and extraction processes of oil and gas, the natural radionuclides ^{235}U, ^{238}U and ^{232}Th as well as the radium-radionuclides (^{223}Ra, ^{226}Ra, ^{228}Ra) and ^{210}Pb etc., are brought to the slurry surfaces and may contain levels of radioactivity above the surface background (Smith et al. 1995a, b).

In the petroleum industry, NORM such as the ones from the ^{232}Th series, as well as ^{40}K are often enhanced as a result of industrial operations, these materials are formally referred to as Technologically Enhanced Naturally Occurring Radioactive Materials (TENORM) (Hrichi et al. 2013). Being relatively insoluble, both uranium and thorium will not be leached and will remain in the oil formation. In contrast, radium is more soluble, and under certain physical and chemical conditions will be leached from the petroleum reservoir rocks to the formation water, which is present together with the oil in the reservoir (Rajaretnam and Spitz 2000; Shawky et al. 2001). As oil is pumped to the surface, water will also come along with it. The produced water extracted with the petroleum contains dissolved mineral salts, some of which may be radioactive, because of the presence of ^{226}Ra and ^{228}Ra and their decay products. The amount of NORM in an oil-producing field generally increases as the amount of produced water pumped with the oil increases. Radium, the predominant radionuclide brought to the surface with the crude oil and produced water, can either stay in solution in the produced water or co-precipitate with barium in the form of complex compounds of sulfates, carbonates, and silicates found in sludge and scale (Gazineu et al. 2005). The formation of these hard, very insoluble precipitates is caused by changes in pressure and temperature as the oil/water mixture is pumped to the surface, the amount of precipitate being dependent on the physical-chemical characteristics of the water (Testa et al. 1994; Rajaretnam and Spitz 2000).

The radioactivity of scale formed in oil and gas production processes is an important issue, particularly from the point of view of radiation protection and has recently received considerable attention (Testa et al. 1994). Although the initial production of oil and gas from a reservoir is typically ‘dry’, formation waters may be increasingly produced along with the oil and gas as the natural pressure within the petroleum-bearing formation decreases. As produced waters are brought to the surface, decreases in temperature and pressure allow solutes contained within the waters to precipitate (Smith 1987). This can result in the formation of hard, extremely insoluble barite scale deposits on the interior surfaces of piping, on casing materials, and on other production equipment. The present study was conducted to review studies conducted on Technically Enhanced Naturally Occurring Radioactive Materials (TE-NORM) waste generated from oil and gas production.

2 Reported Levels of Natural Radionuclides in the Oil Industry

Bou-Rabee et al. (2009) have stated that comprehensive older literature reviews of radium nuclide concentrations in formation and produced water indicated an average radium nuclide concentration in waters in excess of 1.85 Bq/L and exceptionally up to ~1000 Bq/L. As ^{226}Ra originated from the radioactive decay of ^{238}U, while ^{228}Ra from ^{232}Th the $^{226/228}$Ra ratio in the oil-field brines depends on the U/Th ratio of the reservoir rock and ranges from 0.1 to 2.0, but for the most cases its activities are comparable. Typical ranges or average values of the radium radionuclide concentrations in the formation or produced water from different oil fields, including recent data, are listed in Table 1.

Reported levels of the ^{226}Ra and ^{228}Ra activity concentrations observed in the solid scale and sludge are listed in Table 2

3 NORM Action Limits

The action limits for the disposal/control of NORM waste material may be stipulated in National regulations. However, in the absence of National Regulations, the following limits can be utilised and these will provide for compliance with current international practice:

- Materials and waste media such as sludge/scale containing NORM at levels below those listed in Table 3 shall be exempted from the requirements of this procedure.
- Soil shall not have a Radium-226 contamination above 0.185 Bq/(5 pCi/g) above background averaged over any 10 m^2 or unless risk assessment demonstrates an acceptable level of risk.
- Equipment, vessels, and clothing shall be considered 'NORM contaminated' if internal or external surface contamination measures double the radiation background level.

Table 1 Ranges of activity levels in produced water from the oil fields

Country/oil field	Sample	^{226}Ra (Bq/L)	^{228}Ra (Bq/L)	Reference
Egypt	Formation water	5–40	1–59	Shawky et al. (2001)
Algeria	Formation water	5.1–14.8	–	Hamlat et al. (2001)
Norway	Formation water	0.3–10.4	–	Strand and Lysebo (1998)
USA	Produced water	22–30	25–30	Zielinski and Budahn (2007)
Norway	Produced water	9.9–111.2	8.8–60.4	Eriksen et al. (2006)
Norway	Produced water	0.5–16	0.5–21	Norwegian Radiation Protection (2005)
Tunisia	Produced water	0.37–19	–	Hrichi et al. (2013)
US Gulf Coast	Produced water	<0.002–58	0.02–59	Kraemer and Reid (1984)
The Netherlands	Produced water	<2–302	<1–20	Van Hattum et al. (1992)
Norway	Produced water	<DL–10.4	<DL–10.0	Strand et al. (1997)
Brazil	Produced water	<0.01–6.0	<0.05–12.0	Vegueria et al. (2002)
USA, Poland and Austria	Produced water	0.05–191		Fischer (1998)
Norway	Produced water	6–9	<2–17	Røe (1998)
Louisiana, USA	Produced water	<DL–34.4	<DL–34.3	Hamilton et al. (1991)
USA	Produced water	0.1–21.6	0.7–21.7	Hamilton et al. (1991)
US Gulf Coast	Produced water	2–55	2.6–22	Lagera et al. (1999)
Syria	Produced water	50.8–60.3	–	Al-Masri and Suman (2003)
Nigeria	Produced water	3.50–10.81	3.40–9.31	Avwiri et al. (2013)

Table 2 Ranges of activity levels of ^{226}Ra and ^{228}Ra in different scale and sludge samples

Country/oil field	Sample	^{226}Ra (kBq/kg)	^{228}Ra (kBq/kg)	Reference
USA	Scale	15.4–76.1	-	White and Rood (2001)
Brazil	Scale	109.6–3500.0	133.8–2195.0	Gazineu et al. (2005)
Australia	Scale	0.021–0.250	0.048–0.300	Guidelines for NORMs (2002)
Malaysia	Scale	0.114–0.187	0.130–0.206	Omar et al. (2004)
UK	Scale	1–1000	–	Production Forum (1987)
Kazakhstan	Scale	5.10–51.0	0.20–10.0	Kadyrzhanov et al. (2005)
Tunisia	Scale	0.03–1.189	–	Testa et al. (1994)
Brazil	Scale	19.1–323.0	4.21–235.0	Godoy and Cruz (2003)
Norway	Scale	0.3–32.3	0.3–33.5	Lysebo et al. (1996)
Syria	Scale	0.147–1.05	0.043–0.181	Al-Masri and Suman (2003)
Brazil	Scale	77.9–2110	101.5–1550	Gazineu and Hazin (2008)
Member States	Scale	0.0001–15	0.00005–2.8	OGP (2008)
Brazil	Sludge	42.7–167.8	40.5–152.4	Gazineu et al. (2005)
Brazil	Sludge	50.0–168.0	49–52	Godoy and Cruz (2003)
Egypt	Sludge	18	13.25	Shawky et al. (2001)
Norway	Sludge	0.1–4.7	0.1–4.6	Lysebo et al. (1996)
Brazil	Sludge	0.36–367.0	0.25–343.0	Godoy and Cruz (2003)
Syria	Sludge	0.470–1.0	0.359–0.660	Al-Masri and Suman (2003)
Member States	Sludge	0.00005–0.80	0.00005–0.050	OGP (2008)

Table 3 NORM exemption levels (OGP 2008)

Radionuclide	Exemption level (Bq/g)	Exemption level (pCi/g)
^{226}Ra	1.1	30
^{228}Ra	1.1	30
^{210}Pb	0.2	5
^{210}Po	0.2	5
^{238}U	5.5	150
Uranium (nat)	3.0	80

4 Radiological Exposure for Public, Workers, and Environmental Impact of TENORM

Agbalagba et al. (2013) indicated that the mean radiation values to the public in the Nigerian communities of Ughelli East, Kokori, Eriemu, Evwreni, Oweh, Olomoro-Oleh fields are within the 1.0 $mSvy^{-1}$ maximum permissible limit recommended for nonnuclear work environments and the general public, while the values for Otorogu, Ughelli West, Afiesere and Uzere West and East fields exceed this limit. The exposure rates for the host communities range from 0.62 ± 0.16 $mSvy^{-1}$ in Evwreni community to 1.117 ± 0.37 $mSvy^{-1}$ in Otujeremi town. Agbalagba et al. (2013) further observed that proximity to radiation source plays an important role in the radiation impact and distribution. The equivalent dose rate for Otujeremi and Ekakpamre communities exceed the 1.0 $mSvy^{-1}$ maximum permissible limit recommended for non-nuclear work environments and the general public (ICRP 2007). In some of the other communities such as Emeragha, Olomoro and Uzere with equivalent dose rates lying within, but very close to the 1.0 $mSvy^{-1}$ limit, further radiation accumulation may result in the permissible limits for the public being exceeded, thereby resulting in health hazards in these communities.

Rich and Crosby 2013 presented the potential impact of TENORM to the environment, occupational workers, and the general public with the potential health effects of individual radionuclides i.e. cell damage, carcinogenesis, bone marrow damage resulting in anemia among others in a study that analyzed the specific radionuclides present in reserve pits for natural gas mining. The radionuclides present included those from ^{232}Th decay series (^{228}Ra, ^{228}Th, ^{208}Tl) and ^{226}Ra decay series (^{214}Pb, ^{214}Bi, ^{210}Pb) radionuclides.

Naturally Occurring Radioactive Materials (NORM) resulting from the ^{232}Th and ^{238}U-series can be concentrated and accumulated in tubing and surface equipment in the form of scale and sludge as a consequence of physical and chemical processes associated with the oil and gas industry (IAEA 2003; Jonkers et al. 1997).

Testa et al. found that, variations in sulphates and carbonate solubility can give rise to scale formation, which are connected to some physical and chemical factors, e.g., temperature variation, pressure changes, pH-balance, evaporation in the gas extraction pipes and injection of incompatible sea waters. Also, the water re-injected into the reservoirs to maintain the production pressure during the field exploration seemed to be a principal cause for the scale formation (Testa et al. 1998). As a consequence of the physical and chemical processes during the extraction of oil, besides the production water additional scale is obtained. Scale production in gas and oil field equipment is due to precipitation of alkaline earth metal sulphates or carbonates according to the following chemical reactions:

$$Ca^{+2} + CO_3^{-2} \rightarrow CaCO_3$$

$$Sr^{+2} + SO_4^{-2} \rightarrow SrSO_4$$

$$Ba^{+2} + SO_4^{-2} \rightarrow BaSO_4$$

Radium is chemically similar to barium (Ba), strontium (Sr) and calcium (Ca), hence radium co-precipitates with Sr, Ba or Ca scale forming radium sulphate, radium carbonate and—in some cases—radium silicate. Therefore, the observed levels of activity concentrations both in the separated sludge and solid scale are much higher than those observed in the produced water from the oil industry. Many of the physical characteristics of oil formations, high temperature, pressure etc. tend to increase the radionuclide solubility in production fluids. This is due to the complex process for sludge formation. Generally, uranium and thorium are relatively insoluble and remain stationary in the reservoir, while, radium is more soluble and may become mobilized in the produced water phase of the reservoir. It has been estimated that: 25,000 tonnes of TE-NORM contaminated scale are generated each year by the petroleum industry. The available data indicate that, the total radium levels amount to extreme measurements of 15.2 kBq/g scale and 25.9 kBq/g in sludge (Smith et al. 1996).

Numerous studies have been devoted to appraise the real radiation doses and risk for workers in the oil industry. Some results of these investigations are shown in Table 4. The real occupational doses depend on the dose rates and the working time spent during normal activities. The crucial problem in the occupational effective dose evaluation is to assess the so-called occupancy factor. Usually, for typical activities and repair work, this value ranges from 10 to 20 h/year. Calculated on these assumptions the annual effective doses for normal activities in the oil industry should be in the range of up to 2 mSv/year.

Table 4 Exposure rate levels in the oil industry

Country	Reported range (μSv/h)	Reference
Algeria	up to 100	Hamlat et al. (2001)
United Kingdom	10–300	Hamlat et al. (2001)
Egypt	50–100	El Afifi and Awwad (2005)
Congo, Italy, Tunisia	0.1–6	Testa et al. (1994)
USA	up to 300	Jonkers et al. (1997)
Syria	9.2–59	Al-Masri and Suman (2003)

5 The Acceptability of Occupational Risks in Industry

The International Commission on Radiological Protection (ICRP) reviews estimates of radiation risk from every available source, particularly the work of the United Nations Scientific Committee on the Effects of Atomic Radiation (UNSCEAR) and the United States National Academy of Science Committee on the Biological Effects of Ionizing Radiation (BEIR). The reports of the ICRP go further than these sources, in that the ICRP recommends permissible exposures for workers while the other bodies merely estimate the risks associated with radiation exposure (Radiation Health & Safety Advisory Council 2005).

The ICRP (2007) believes that any exposure to ionizing radiation may be potentially harmful to health, and advocates three fundamental principles for managing radiation exposures:

- **JUSTIFICATION**—No activity involving ionizing radiation for any purpose can be justified unless it is possible to demonstrate that it will lead to a positive net benefit.
- **OPTIMIZATION**—All exposures shall be kept as low as reasonably achievable, economic and social factors being taken into consideration; (the ALARA principle).
- **LIMITATION**—The maximum acceptable occupational exposure of any individual must not involve a radiation risk to that individual greater than the risk that arises in working in what is generally regarded as a "safe" industry.

The ICRP (2007) recognizes that everyone is subject to a significant background radiation exposure. However, even smaller-than-background doses from occupational practices are unjustifiable if there is no associated benefit, or they can be readily avoided.

6 Health Effects of Ionizing Radiations

Ionizations, received in sufficient quantities over a period of time, can result in tissue damage and disruption of cellular function at the molecular level. Of particular interest is their effect on deoxyribonucleic acids (DNA). The dose delivered to tissue from ionizing radiation can either be acute (the energy from the radiation is absorbed over a few hours or days) or chronic (the energy is absorbed over a longer period of months, years, or over a lifetime). High doses of ionizing radiation can lead to acute effects, such as skin burns, hair loss, birth defects, illness, cancer, and death. The basic principle of toxicology, "the dose determines poison," applies to the toxicology of ionizing radiation as well as to all other branches of toxicology. In the case of threshold effects ("deterministic effects" in the language of radiation toxicology), such as skin burns, hair loss, sterility, nausea, and cataracts, a certain minimum dose (the threshold dose), usually on the order of hundreds or thousands of rad, must be exceeded in order for the effect to be expressed. An increase in the size of the dose above the threshold dose will increase the severity of the effect.

Exposure to (TE)NORM will not result in acute and severe effects similar to those effects associated with exposure to high radiation levels, it can however result in delayed effects such as the development of certain forms of cancer such as leukemia, and cancers of the lung, stomach, esophagus, bone, thyroid, and the brain and nervous system. For cancer induction, increasing the radiation dose does not increase the severity of the cancer; instead it increases the chance of cancer induction. In the case of carcinogens generally, whether chemical or radiological, safety standards are based on a postulated zero threshold (i.e., any increment of carcinogen, no matter how small, is assumed to carry with it a corresponding increase in the chance of causing cancer). Increasing the size of the dose increases the probability of inducing a cancer with that carcinogen. Cancers that are, in fact, caused by radiation are completely indistinguishable from those that seem to occur spontaneously or are caused by other known or suspected carcinogens.

All biological damage effects begin with the consequence of radiation interactions with the atoms forming the cells resulting in deterministic or stochastic effect. There are two mechanisms by which the radiation ultimately affects cells. These two mechanisms are direct or indirect effect/action (ATSDR 1999).

7 Recommended Radiation Dose Limits

It is the recommendation of the Radiation Protection Institute of Ghana that the annual incremental effective dose to persons exposed to NORM as the result of a work practice be limited to the values given below (Table 5).

In the 2007 Recommendations of the International Commission on Radiological Protection (ICRP Publication No. 103), the recommended dose limits for exposure associated with practice are as follows:

Table 5 Dose limits used at the National Radioactive Waste Management Centre, Ghana (RPB 1995)

Type of limits	Occupational exposure limits	Public exposure limits
Effective dose	20 mSv/year, averaged over 5 consecutive years; 50 mSv in a single year	5 mSv in a year
Annual equivalent dose		
Eye lens	150 mSv in a year	15 mSv in a year
Skin	500 mSv in a year	50 mSv in a year
Hands and feet (Extremeties)	500 mSv in a year	–

Table 6 ICRP recommended dose limits (ICRP 2007)

Application	Persons employed in radiation work	Public
Whole body effective dose	20 mSv/year, averaged over defined in periods of 5 years.	1 mSv/year
	Not exceed 50 mSv in any single year	
Annual equivalent dose in		
the lens of the eye	150 mSv	15 mSv
the skin	500 mSv	50 mSv
the hands and feet	500 mSv	–

1. The limits apply to the sum of the relevant doses from external exposure in the specified period and the 50-year committed dose (to age 70 years for children) from intakes in the same period.
2. Additional restrictions apply to the occupational exposure of pregnant women.
3. In special circumstances, a higher value of effective dose could be allowed in a single year, provided that the average over 5 years does not exceed 1 mSv/year.
4. The limitation on the effective dose provides sufficient protection for the skin against stochastic effects. An additional limit is needed for localized exposures in order to prevent deterministic effects.

The U.S. Nuclear Regulatory Commission set a radiation exposure limit of 5 mSv for pregnant working women over the full gestational period (USNRC 1991). For the critical gestational period of 8–15 weeks ATSDR believes that the acute Minimal Risk Level of 4 mSv is consistent with the NRC limit and could be applied to either acute (0–14 days) or intermediate (15–365 days) exposure periods.

8 Recommendations

A literature survey on the radiological impact of oil and gas was carried out. From the obtained results, it is found that the average activity concentration of ^{226}Ra and ^{228}Ra are above exemption level as set by IAEA (2006) and occupational annual effective dose due to direct gamma exposure and dust inhalation are within the international recommendations. It is recommended that special attention should be focused on the workers who are in charge of maintaining and repairing equipment in order to control factors that could affect their exposure to radiation. Specifically, they should not be allowed to enter into the tanks to remove oil residues and scale if good ventilation conditions are not provided. Besides, all cleaning operations that require scraping or similar procedures for scale removal must be performed in wet conditions, in order to minimize the inhalation or ingestion of radioactive dust. The use of protective masks is also required during all cleaning operations. The data obtained must be helpful in setting a national screening level of exposure to NORM in different activities, and developing national control regulation dealing with

NORM industries. The current practise of disposal of solid radioactive scale or sludge from the oil industry to marine or to sanitary landfills should be carefully examined and is generally not recommended. A case has therefore been made for the Ghana Radiation Protection Institute to consider repositories situate within underground rock formations as the safest way for the final storage destination for the majority of solid scale waste with specific activities in the range of 10–100 kBq/kg. Moreover, there is an urgent need to establish clear rules and national clearance or exemption levels for selected radionuclides for discharge to deal with radioactive waste especially from the oil industry.

9 Conclusion

1. Some increased NORM concentration in the scale, sludge, produced water and formation samples were reported. The major polluters are ^{226}Ra and ^{228}Ra with the maximum specific activity of 3500 kBq/kg and 2195 kBq/kg respectively.
2. On the basis of the review of available data describing the occurrence of NORM in the oil and gas industry, preliminary conclusion can be drawn regarding the need for additional research.

10 Summary

Oil and gas production processing operations have been known to produce a large amount of naturally occurring radioactive materials (NORMs) at elevated concentrations as by-product waste streams. A comprehensive literature review was carried out to assess the activity concentration of ^{226}Ra, ^{228}Ra in scales, sludge, produced water and formation water and the subsequent radiological impact of this industry on workers and the public. The observed specific activities of ^{226}Ra are in the ranges up to 3500 kBq/kg, 367 kBq/kg, 1200 Bq/kg and 40 Bq/kg for scale, sludge, produced water and formation water respectively. ^{228}Ra has up to 2195 kBq/kg, 343 kBq/kg, 180 Bq/kg, and 59 Bq/kg for solid scale and sludge, produced water and formation water respectively. The activities of both radionuclides exceed the exemption level of 10 kBq/kg recommended by IAEA safety standards. This means that TENORM wastes from the oil industry may generate radiation exposure levels which require attention and continuous monitoring during some routine operations in this industry. This exposure is mostly caused by external γ-radiation coming from the ^{228}Ra radionuclide and its progenies. The field of radiation protection and corresponding national and international regulations has evolved to ensure safety in the use of radioactive materials.

References

Agbalagba EO, Avwiri GO, Chadumoren YE (2013) J Appl Sci Environ Manage 17(2):279–288

Agency for Toxic Substances and Disease Registry (1999) Toxicological profile for ionizing radiation, toxic substances portal – ionizing radiation, http://www.atsdr.cdc.gov/toxprofiles. Accessed on Jul 25, 2015

Al-Masri MS, Suman H (2003) NORM waste management in oil and gas industry: the Syrian experience. J Radioanal Nucl Chem 256(1):159–162

IAEA (International Atomic Energy Agency) (2003) Radiation and Waste Safety in the Oil and Gas Industry. IAEA-Safety Report No. 34

Avwiri GO, Esi E, Agbalagba EO (2013) Gamma spectrometry analysis of produced water from selected flow stations in Delta State, Nigeria. Int J Environ Monitor Anal 1(5):167–174

Azeri-Chirag-Gunashli (2004) Full field development phase 3, Environmental and socio-economic impact assessment 31648 – 046, Appendix 10

Bou-Rabee F, Al-Zamel AZ, Al-Fares RA, Bem H (2009) Technologically enhanced naturally occurring radioactive materials in the oil industry (TENORM). A review. NUKLEONIKA 54 (1):3–9

Doyi I, Oppon OC, Glover ET, Gbeddy G, Kokroko W (2013) Assessment of occupational radiation exposure in underground artisanal gold mines in Tongo, Upper east region of Ghana. J Environ Radioact 126:77–82

El Afifi EM, Awwad NS (2005) Characterization of the TENORM waste associated with oil and natural gas production in Abu Rudeis, Egypt. J Environ Radioact 82:7–19

Eriksen DO, Sidhu R, Strälberg E (2006) Radionuclides in produced water from Norwegian oil and gas installations-concentrations and bioavailability. Czechoslovak J Phys 56:D43–D48

Fischer RS (1998) Geologic and geochemical controls on naturally occurring radioactive materials (NORM) in produced water from oil, gas, and geothermal operations. Environ Geosci 5 (3):139–150

Gazineu MHP, Hazin CA (2008) Radium and potassium-40 in solid wastes from the oil industry. Appl Radiat Isot 60:90–94

Gazineu MHP, de Araujo AA, Brandao YB, Hazin CA, Godoy JM (2005) Radioactivity concentration in liquid and solid phases of scale and sludge generated in the petroleum industry. J Environ Radioact 81:47–54

Godoy JM, Cruz RP (2003) ^{226}Ra and ^{228}Ra in scale and sludge samples and their correlation with the chemical composition. J Environ Radioact 70:199–206

Guidelines for naturally occurring radioactive materials (2002) Australian Petroleum Production & Exploration Associated Ltd. Report ABN 44000292773, March 2002, Canberra

Hamilton LD, Meinhold AF, Nagy J (1991) Produced water radionuclide hazard/risk assessment – phase 1. Prepared for the American Petroleum Institute. BNL 47403. Brookhaven National Laboratory

Hamlat MS, Djeffal S, Kadi H (2001) Assessment of radiation exposures from naturally occurring radioactive materials in the oil and gas industry. Appl Radiat Isot 55:141–146

Hrichi H, Baccouche S, Belgaied J (2013) Evaluation of radiological impacts of tenorm in the Tunisian petroleum industry. J Environ Radioact 115(2013):107–113

IAEA (2006) Regulatory and management approaches for the control of environmental residues containing naturally occurring radioactive material NORM – Proceedings of a Technical Meeting held in Vienna, 6–10 December 2004. TECDOC-1484. IAEA, Vienna

ICRP (2007) Annals of the ICRP Publication 103, The 2007 Recommendations of the International Commission on Radiological Protection, pp 103–122

Jonkers G, Hartog FA, Knaepen WAI, Lancee PFJ (1997) Characterization of NORM in the oil and gas production (E&P) industry. In: Proceedings of the International Symposium on Radiological Problems with Natural Radioactivity in the Non-nuclear Industry. Amsterdam, September 8–10, pp. 23–47

Kadyrzhanov KK, Tuleushev AZ, Marabaev ZN (2005) Radioactive components of scales at the inner surface of pipes in oil fields of Kazakhstan. J Radioanal Nucl Chem 264:413–416

Kraemer TF, Reid DF (1984) The occurrence and behaviour of radium in saline formation water of the U.S. Gulf Coast Region. Isot Geosci 2:153–174

Lagera L, Hart A, Graham B (1999) Radionuclides in oil and gas operational discharges and environmental samples associated with offshore oil and gas production facilities. I: Continental Shelf Associates, Inc. Radionuclides, metals, and hydrocarbons in oil and gas operational discharges and environmental samples associated with offshore production facilities on the Texas/Louisiana continental shelf with an environmental assessment of metals and hydrocarbons. Prepared for U.S. Department of Energy. (DOE/MT/92001-23). Tulsa, OK (US): National Petroleum Technology Office, 1999: chapter 6

Lysebo I, Birovljev A, Strand T (1996) NORM in oil production, occupational doses and environmental aspects. In: Proceedings of the 11th Congress of the Nordic Radiation Protection Society, Reykjavik, August 26–30, http://www.gr.is/nsfs_eng.html. Accessed on Jan 28, 2014

Norwegian Radiation Protection Authority (2005) Natural radioactivity in produced water from the Norwegian oil and gas industry in 2003, Report no 2. NRPA, Østeras

OGP (2008) Guidelines for the management of naturally occurring radioactive materials in the oil and gas industry, International Association of Oil & Gas Producers, Report No. 412

Omar M, Ali HM, Abu MP (2004) Distribution of radium in oil and gas industry wastes from Malaysia. Appl Radiat Isot 60:779–782

Pasocha AS (1997) Naturally occurring radioactive materials (NORM) and petroleum origin. Appl Radiat Isot 48:1391–1396

Exploration and Production Forum (1987) Low specific activity scale origin treatment and disposal. Report no 6.6/127. London, pp 25–38, Final version 17 August 2005

Radiation Health & Safety Advisory Council for the CEO of ARPANSA, (2005) Naturally-Occurring Radioactive Material (NORM) in Australia: Issues for Discussion

Rajaretnam G, Spitz HB (2000) Effect of leachability on environmental risk assessment for naturally occurring radioactive materials in petroleum oil fields. Health Phys 78(2):191–198

Rich AL, Crosby EC (2013) Analysis of reserve pit sludge from unconventional natural gas hydraulic fracturing and drilling operations for the presence of Technologically Enhanced Naturally Occurring Radioactive Material (TENORM)

Røe TI (1998) Produced water discharges to the North Sea: a study of bioavailability of organic produced water compounds to marine organisms. PhD thesis. Trondheim: Norwegian University of Science and Technology, Faculty of Chemistry and Biology

RPB (1995) Dose limits, radiation protection and safety guide No. GRPB-G3, Radiation Protection Board, Ghana Atomic Energy Commission

Shawky S, Amer H, Nada AA, El-Maksoud TMA, Ibrahiem NM (2001) Characteristics of NORM in the oil industry from Eastern and Western deserts of Egypt. Appl Radiat Isot 55:135–139

Smith AL (1987) Radioactive scale formation. J Petrol Tech 39:697–706

Smith GE, Fitzgibbon T, Karp S (1995a) In Proceedings of Exploration and Production Environmental Conference. Houston, Texas, USA, March 7–29, p 181

Smith KP, Blunt DL, William GP, Tebes LL (1995b) In Proceedings of Exploration and Production Environmental Conference. Houston, Texas, USA, March 27–29, p 231

Smith KP, Blunt DL, Williams GP, Tebes CL (1996) 44. Radiological dose assessment related to management of naturally occurring radioactive materials generated by the petroleum industry. Report ANL/EAD-2m. Argonne National Laboratory, USA

Snavely ES (1989) Radionuclides in produced water - a literature review. American Petroleum Institute Publication No. 5404, Washington, DC, pp 1–85

Strand T, Lysebo I (1998) NORM in oil production activity levels and occupational doses. In: Proc of the 2nd Int Symp on the Treatment of Naturally Occurring Radioactive Materials NORM II, 10–13 November 1998, Klefeld, Germany, pp 137–141

Strand T, Lysebo I, Kristensen D, Birovljev A (1997) Deposits of naturally occurring radioactivity in production of oil and natural gas. NRPA report 1997:1. Østerås, Norwegian Radiation Protection Authority, 1997. (In Norwegian)

Testa C, Desideri D, Meli MA, Roselli C, Bassignani A, Colombo G, Fantoni RF (1994) Radiation protection and radioactive scales in oil and gas production. Health Phys 71:34–38

Testa, C, Desideri D, Guerra F, Meli MA, Roselli C (1998) Proc. of the Int. Solv. Extraction Symp., Moscow, Russia, June 21–27, 416

USNRC. 1991. Occupational dose limits. U.S. Nuclear Regulatory Commission. Code of Federal Regulations. 10 CFR 20, Subpart C

Van Hattum B, Cofino WP, Feenstra JF (1992) Environmental aspects of produced water discharges from oil and gas production on the Dutch continental shelf. Amsterdam: Vrije University, Institute for Environmental Studies, 1992

Vegueria JSF, Godoy JM, Miekeley N (2002) Environmental impact studies of barium and radium discharges by produced waters from the "Bacia de Campos" oil field offshore platforms, Brazil. J Environ Radioact 62:23–38

White GJ, Rood AS (2001) Radon emanation from NORM-contaminated pipe scale and soil at petroleum industry sites. J Environ Radioact 54:401–413

Zielinski RA, Budahn JR (2007) Mode of occurrence and environmental mobility of oil-field radioactive material at US Geological Survey research site B. Appl Geochem 22:2125–2137

Index

P. de Voogt (ed.), *Reviews of Environmental Contamination and Toxicology, Volume 238*, Reviews of Environmental Contamination and Toxicology 238,
DOI 10.1007/978-3-319-30791-6

The manufacturer's authorised representative in the EU is Springer Nature Customer Service Centre GmbH, Europaplatz 3, 69115 Heidelberg, Germany. If you have any concerns regarding our products, please contact ProductSafety@springernature.com

Printed and bound by CPI Group (UK) Ltd, Croydon, CR0 4YY
28/07/2026
02179219-0018